Springer Remote Sensing/Photogrammetry

The Springer Remote Sensing/Photogrammetry series seeks to publish a broad portfolio of scientific books, aiming at researchers, students, and everyone interested in the broad field of geospatial science and technologies. The series includes peer-reviewed monographs, edited volumes, textbooks, and conference proceedings. It covers the entire area of Remote Sensing, including, but not limited to, land, ocean, atmospheric science and meteorology, geophysics and tectonics, hydrology and water resources management, earth resources, geography and land information, image processing and analysis, satellite imagery, global positioning systems, archaeological investigations, and geomorphological surveying.

Fatwa Ramdani

Exploring the Earth with QGIS

A Guide to Using Satellite Imagery at Its Full Potential

Springer

Fatwa Ramdani
University of Tsukuba
Tsukuba, Ibaraki, Japan

ISSN 2198-0721 ISSN 2198-073X (electronic)
Springer Remote Sensing/Photogrammetry
ISBN 978-3-031-46041-8 ISBN 978-3-031-46042-5 (eBook)
https://doi.org/10.1007/978-3-031-46042-5

This Springer imprint is published by the registered company Springer Nature Switzerland AG
The registered company address is: Gewerbestrasse 11, 6330 Cham, Switzerland

Paper in this product is recyclable.

For Putri Setiani, Alidrisi Sorai Ramdani,
and Albiruni Keiji Ramdani

Preface

In recent years, the field of earth remote sensing has exploded with the development of new satellite technologies and the availability of vast amounts of high-resolution imagery. As a result, there is a growing demand for GIS professionals who are skilled in the use of remote sensing techniques to analyze and interpret these data sets.

With the rapid advancements in satellite technology, earth remote sensing has become an indispensable tool for a wide range of applications, from environmental monitoring to disaster management. However, the vast amount of data generated by these satellites can be overwhelming for many researchers and practitioners who lack the necessary skills and tools to process and analyze them.

This is where QGIS comes in. QGIS is a powerful, open-source Geographic Information System (GIS) software that has been extensively used in the field of earth remote sensing. It provides a user-friendly platform for processing, analyzing, and visualizing satellite imagery, enabling users to unlock the full potential of earth remote sensing data.

QGIS is an open-source GIS software that has quickly become popular for earth remote sensing applications due to its intuitive interface and powerful analysis tools. In this book, I have covered the entire process from acquiring and preprocessing data to conducting advanced analysis, interpretation of the results, and provided a comprehensive guide to using QGIS for unlocking the potential of satellite imageries. The book is divided into 10 main chapters:

1. Introduction to QGIS
2. QGIS for Remote Sensing
3. Data Acquiring
4. Data Preprocessing
5. Data Management
6. Data Visualization
7. Data Analysis
8. Integration of Remote Sensing Data with Other Spatial Datasets
9. Integration of QGIS with Google Earth Engine (GEE)
10. Case Studies and Examples of Earth Remote Sensing Applications Using QGIS

This book is intended for GIS professionals, students, and researchers who are interested in using QGIS for earth remote sensing applications. It will be particularly useful for those who are new to QGIS or who want to expand their knowledge of the software for remote sensing purposes. However, there are still many things that are not covered in this book. Readers with advanced knowledge about remote sensing and satellite imagery processing are suggested to look for more references.

Finally, I would like to thank my wife and family for always supporting me in spreading the science and knowledge of geospatial computing.

Tsukuba City, Ibaraki, Japan
2023

Fatwa Ramdani

Contents

List of Figures

List of Tables

Chapter 1
Introduction to QGIS

Quantum Geographic Information System (QGIS) is a free and open-source GIS software that is used for creating, editing, analyzing, and visualizing spatial data. QGIS provides a user-friendly interface. It is capable of handling a wide range of raster and vector data formats, including satellite and aerial imagery, elevation data, and land cover data. It is widely used in various industries including government, academia, and private sectors. One of the many functions of QGIS is its ability to process remote sensing data.

Remote sensing refers to the acquisition of information about an object or phenomenon without physically being in contact with it. This is done through the use of sensors that capture data from a distance, such as satellite images or aerial photographs. Remote sensing data can provide valuable insights into various environmental and societal issues, such as climate change, land use, and natural resource management.

QGIS plays an important role in Earth remote sensing by providing the tools necessary to analyze and visualize remote sensing data. QGIS can be used to preprocess and enhance remote sensing data, such as applying image enhancements, filtering, mosaicking, and clipping.

In addition, QGIS can be used to perform a range of spatial analysis tasks on remote sensing data, including land cover classification, change detection, and spatial modeling. QGIS also provides tools for visualizing remote sensing data in 2D and 3D, such as creating fly-through animations of 3D terrain models. QGIS is a powerful tool for working with remote sensing data and plays a critical role in many Earth remote sensing applications.

F. Ramdani, *Exploring the Earth with QGIS*, Springer Remote Sensing/Photogrammetry, https://doi.org/10.1007/978-3-031-46042-5_1

1.1 Download and Install QGIS

Before you are working on remote sensing analysis, you need to download and install QGIS. Following are the steps to download and install QGIS:

- Go to the QGIS website at https://www.qgis.org/en/site/ and select the "Download" button on the top menu bar.
- Select your operating system. You can choose between Windows, Mac, and Linux.
- Choose the latest version of QGIS for your operating system and select the appropriate download link.
- Once the download is complete, run the installer and follow the installation wizard prompts to install QGIS.
- After installation, launch QGIS and start using it for your GIS tasks.

It's important to note that during the installation process, you may be asked to install additional dependencies or software packages that QGIS requires. Be sure to read the installation prompts carefully and follow the instructions to complete the installation successfully.

1.2 The Differences Between QGIS and Other GIS Software

There is a wide range of GIS software available today, including both commercial and free open-source options. However, while some software may have limited functionality, others require training before they can be effectively used.

To highlight the features of QGIS compared to other GIS software currently available, please refer to Table 1.1.

Table 1.1 Comparison of QGIS feature with other GIS software

Feature	QGIS	Other GIS software
Cost	Free and open-source	Commercial and/or proprietary
User interface	Intuitive and user-friendly	Varied, some are more complex and require training
Data formats	Supports a wide range of formats	Limited support for some formats
Plugins and customization	Highly customizable through plugins	Limited customization options
Platform support	Runs on multiple platforms (Windows, MacOS, Linux)	Platform dependent
Spatial analysis tools	Robust and extensive	Varies depending on software
Community support	Large and active community of users and developers	Varies depending on software
3D visualization	Limited support for 3D visualization	Some software offer advanced 3D visualization tools
Integration with other tools	Good integration with other open-source tools	Limited integration options
Performance	May be slower for large datasets	Varies depending on software

1.3 QGIS Appearance

When you open QGIS, you'll see a menu bar at the top of the screen, which includes several drop-down menus such as "Project," "Layer," "View," "Settings," "Plugins," and "Help." Each of these menus contains various tools and options for working with geospatial data.

The main canvas in QGIS is where you view and work with your map. This canvas displays your data layers, and you can interact with them by zooming in and out, panning, and selecting features. You can also add new layers to the canvas by clicking on the "Add Layer" button in the Layers panel, which is usually located on the left-hand side of the screen.

On the left panel, there are "Browser" and "Layers" panels where you can manage to locate the file and open it. Then you can arrange these files in the Layers panel. In principle, the point, line, and polygon will be located above the raster layer.

When you already have some projects saved, QGIS will show these projects in the "Recent Projects" window in the center. Sometimes it will also show some news related to QGIS as shown in Fig. 1.1.

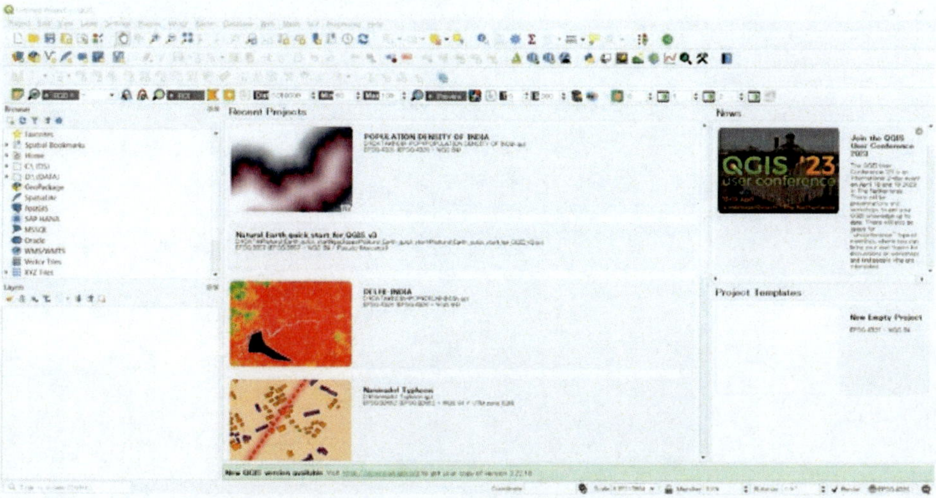

Fig. 1.1 QGIS user interface

Chapter 2
QGIS Plugins for Remote Sensing

One of the most valuable features of QGIS is its plugin system, which allows users to add new functionality and extend the software's capabilities. A plugin is a piece of software that extends the functionality of a larger software application. In the context of QGIS, plugins are small programs that can be installed to add new features to the software. There are hundreds of plugins available for QGIS, and they can be used to add new tools, automate workflows, and enhance the user experience.

Plugins can be installed from within QGIS, and they are accessible through the Plugins > Manage and Install Plugin. The Plugin Manager (Fig. 2.1) allows users to browse available plugins, install and uninstall plugins, and configure plugin settings.

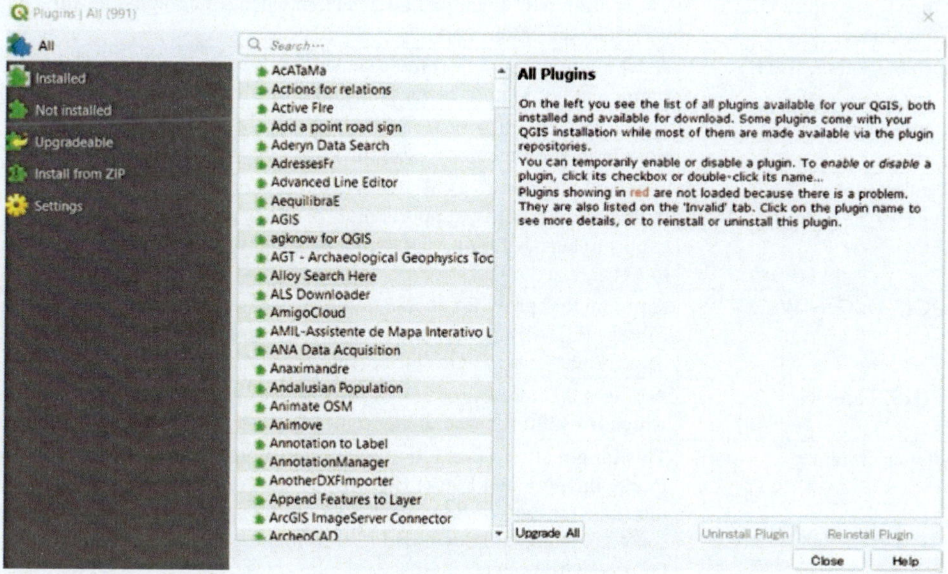

Fig. 2.1 Plugin manager

F. Ramdani, *Exploring the Earth with QGIS*, Springer Remote Sensing/Photogrammetry, https://doi.org/10.1007/978-3-031-46042-5_2

One of the great things about QGIS plugins is that they are developed by a community of users and developers. This means that there are plugins available for a wide range of tasks, and new plugins are being developed all the time. Some plugins are developed by professional developers, while others are created by individual users who have identified a need and developed a solution.

2.1 QGIS Plugin for Remote Sensing Analysis

Plugins can be used to add new functionality to QGIS in a wide range of areas. For example, there are plugins that add support for new data formats, such as optical remote sensing data or Light Detection and Ranging (LiDAR) data. There are also plugins that add new analysis tools, such as tools for calculating slope or aspect from Digital Elevation Model (DEM) data. Additionally, there are plugins that add support for land use and land cover classification. Table 2.1 summarized the plugins that are extensively used for remote sensing analysis.

Table 2.1 List of QGIS plugins for remote sensing analysis

Plugin name	Description
Dzetsaka: classification tool	Plugin for semi-automatic classification with Gaussian Mixture Model, Random Forest, and SVM classifiers
LecoS	A tool for landscape ecology analysis, including habitat fragmentation analysis and land cover change detection
OpenTopography DEM Downloader	This plugin will download DEMs from OpenTopography.org for user-selected extent only
Orfeo Toolbox (OTB)	A set of algorithms for remote sensing processing, including image segmentation, feature extraction, and classification
Point Sampling Tool	A tool for extracting raster values at point locations, such as ground truth points for accuracy assessment
Profile Tool	This tool plots profile lines from raster layers or point vector layers with elevation fields, supports multiple lines as well as graph export to svg, pdf, png, or csv file, and supports 3D polyline export to dxf
QGIS-OSGeo4W	A plugin that provides access to the Geospatial Data Abstraction Library (GDAL) and related tools for working with remote sensing data formats such as GeoTIFF, HDF5, and netCDF
QGIS2ThreeJS	A plugin that allows for 3D visualization of raster and vector data, which is useful for visualizing remotely sensed data in a 3D context
Planet_Explorer	This plugin allows users to efficiently explore, access, and acquire Planet imagery and Planet Basemaps. This plugin offers an interface for discovering imagery, enabling users to seamlessly search, stream, and download Planet imagery directly within the QGIS environment.
Processing Saga NextGen Provider	A processing provider for SAGA 7.2
Semi-Automatic classification Plugin (SCP)	A tool for image classification and analysis, including supervised and unsupervised classification, and feature extraction

Table 2.1 (continued)

Plugin name	Description
SentinelHub	A plugin that allows you to access and download Sentinel satellite imagery directly within QGIS
Serval	A plugin that provides tools for radar image processing, including calibration, speckle filtering, and terrain correction
Temporal/Spectral Profile Tool	A QGIS plugin for interactive plotting of temporal or spectral information stored in multi-band rasters. Based on the Profile tool plugin
TimeManager	A plugin that allows you to create animations and visualizations of temporal data, including remote sensing time series data
WhiteboxTool (WBT)	WBT is an advanced geospatial software package and a data analysis platform with functionality in GIS, remote sensing, LiDAR data processing, DEM analysis, spatial hydrology, and stream network analysis
Zonal Statistics	A plugin that allows you to calculate statistics for raster layers within specified zones or polygons

Note: This is not an exhaustive list and other plugins may be available for remote sensing analysis in QGIS

To find and install a plugin in QGIS go to Plugins > Manage and Install Plugins. The list of all available QGIS plugins, including both installed and downloadable ones, can be found on the left. While some plugins come pre-installed with QGIS, the majority can be accessed via the plugin repositories. To temporarily enable or disable a plugin, simply click its checkbox or double-click its name. Any plugins that are not functioning properly will appear in red and can be found on the "Invalid" tab. Click on a plugin's name (Fig. 2.2) to view additional details or to reinstall or uninstall the plugin.

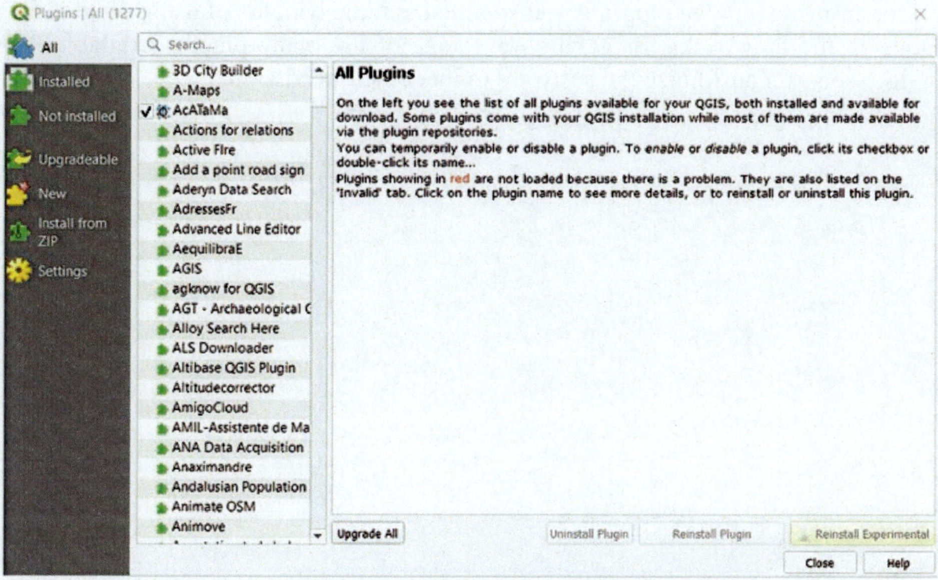

Fig. 2.2 Select the plugin

2.2 Advantages and Disadvantages of Plugins

One of the great things about QGIS plugins is that they are designed to be modular and flexible. This means that users can customize the software to meet their specific needs, and they can easily add or remove plugins as their requirements change. For example, a user might install a plugin to add support for a specific data format and then remove the plugin once they no longer need that functionality.

Another advantage of QGIS plugins is that they are often open source, which means that users can access the source code and modify the plugin to meet their specific needs. This can be especially useful for advanced users who want to extend the functionality of a plugin or create their own custom plugins.

QGIS plugins are also designed to work seamlessly with the core QGIS software. This means that users can use plugins to automate workflows or add new functionality without having to switch between different software applications. Additionally, plugins are often designed to work with other plugins, which means that users can build complex workflows by combining different plugins together.

One of the challenges of using QGIS plugins is that they can be difficult to find and install. While the Plugin Manager makes it easy to browse and install plugins, there are often many different plugins available for a specific task, and it can be hard to determine which plugin is the best fit for a specific workflow or project. Additionally, some plugins may not be well-maintained or may not work with the latest version of QGIS.

In conclusion, plugins are a powerful and flexible way to extend the functionality of QGIS. They can be used to automate workflows, add new tools, and enhance the user experience. There are hundreds of plugins available for QGIS, and they are developed by a community of users and developers. While there are some challenges associated with using plugins, such as finding the right plugin for a specific task or ensuring that the plugin is well-maintained, the benefits of using plugins far outweigh the drawbacks. By taking advantage of the many plugins available for QGIS, users can customize the software to meet their needs.

Chapter 3
Data Acquiring

QGIS, a powerful and versatile geospatial software, empowers us to seamlessly process and analyze a wide array of Earth remote sensing satellite imageries sourced from various providers and platforms. With QGIS, we gain the capability to effortlessly harness the valuable insights and information embedded within these diverse sources of geospatial data, enabling us to make informed decisions and derive meaningful conclusions. These sources encompass an extensive range of satellites, including but not limited to high-resolution optical sensors such as Landsat and Sentinel, as well as synthetic aperture radar (SAR) satellites such as ALOS. By harnessing the advanced functionalities of QGIS, we can effortlessly integrate, visualize, and manipulate these heterogeneous satellite datasets, unlocking a whole new realm of possibilities for environmental monitoring, land cover classification, disaster management, urban planning, and countless other applications in the field of geospatial analysis.

3.1 ALOS

The Advanced Land Observing Satellite (ALOS) is a Japanese Earth-imaging satellite from JAXA (Fig. 3.1). The PALSAR-2 aboard the ALOS-2 is a SAR, which emits microwaves and receives the reflection from the ground to acquire information. Since it does not need other sources of light such as the sun, SAR has the advantage of providing satellite images regardless of day or night. You can download the data from: https://www.eorc.jaxa.jp/ALOS/en/dataset/fnf_e.htm

F. Ramdani, *Exploring the Earth with QGIS*, Springer Remote Sensing/Photogrammetry, https://doi.org/10.1007/978-3-031-46042-5_3

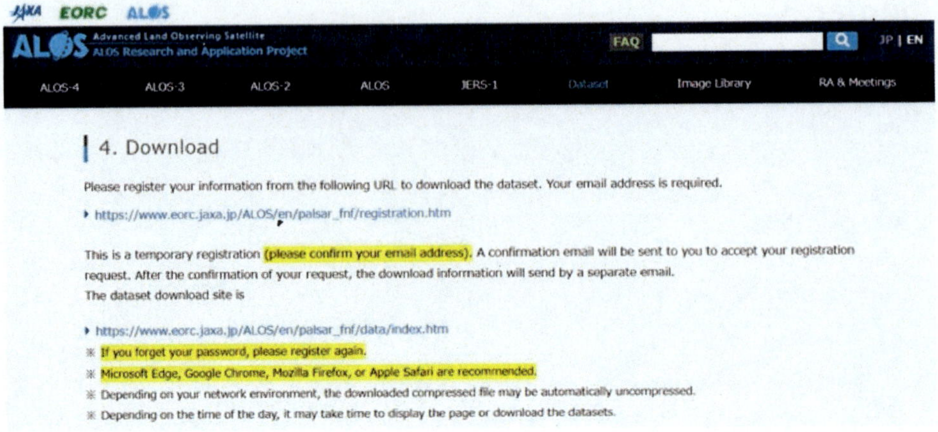

Fig. 3.1 ALOS website

3.2 Landsat

Landsat is a joint mission between NASA and the United States Geological Survey (USGS) that provides global coverage of Earth's land surface. Landsat data can be accessed through the USGS Earth Explorer (Fig. 3.2) website (https://earthexplorer.usgs.gov/).

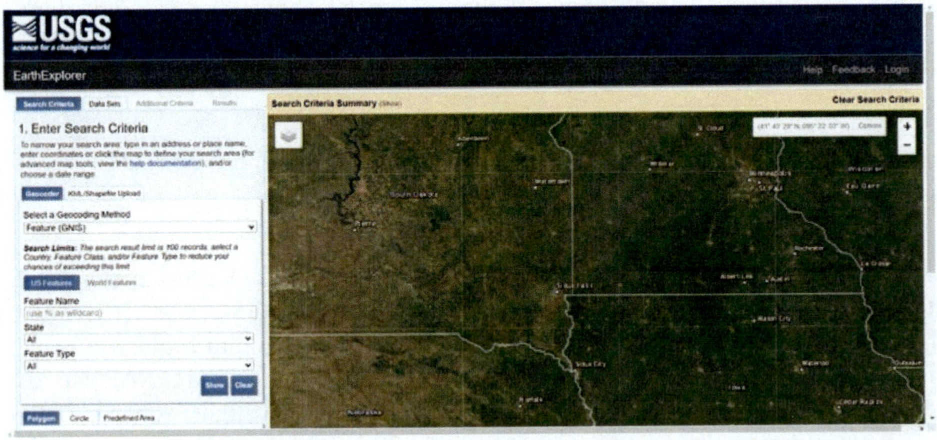

Fig. 3.2 USGS Earth Explorer

3.3 Sentinel

Sentinel is a family of European Earth observation satellites developed by the European Space Agency (ESA). Sentinel data can be accessed through the Copernicus Open Access Hub (Fig. 3.3) (https://scihub.copernicus.eu/dhus/#/home). There are many types of Sentinel sensors such as Sentinel-1 for SAR data, Sentinel-2 and 3 for optical data, and Sentinel-5 for pollutant datasets such as SO2, NO2, etc. Especially for Sentinel-5 is accessible for download at https://s5phub.copernicus.eu/dhus/#/home and you can use **username: s5pguest**, **and password: s5pguest** to log in.

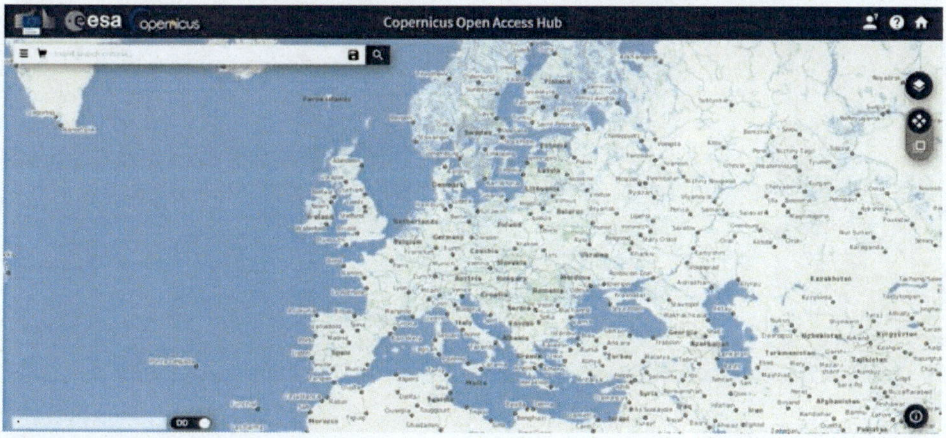

Fig. 3.3 Copernicus Open Access Hub

3.4 MODIS

The Moderate Resolution Imaging Spectroradiometer (MODIS) is a key instrument aboard two NASA Earth Observing System (EOS) satellites that provides daily global coverage. MODIS data can be accessed through the NASA Earthdata Search website (https://search.earthdata.nasa.gov/) (Fig. 3.4).

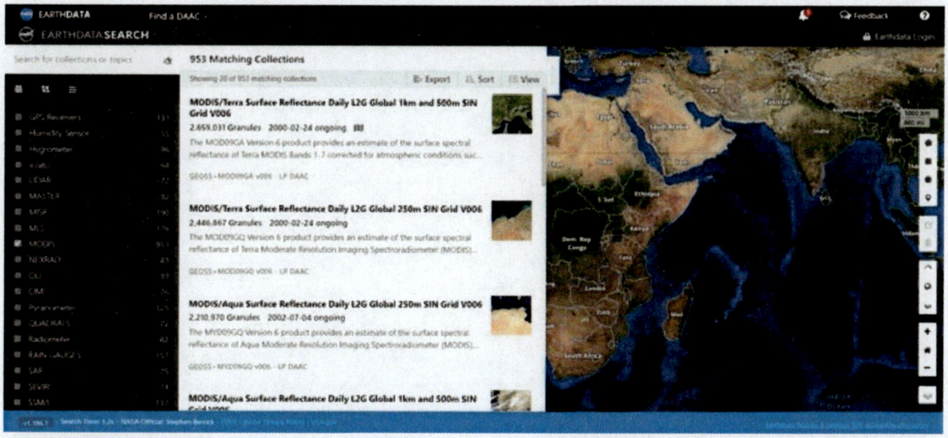

Fig. 3.4 NASA Earthdata Search website

3.5 ASTER

The Advanced Spaceborne Thermal Emission and Reflection Radiometer (ASTER) is a joint mission between NASA and the Japanese Ministry of Economy, Trade, and Industry (METI) that provides high-resolution images of Earth's surface. ASTER data can be accessed through the NASA Earthdata Search website (https://search.earthdata.nasa.gov/) (Fig. 3.5).

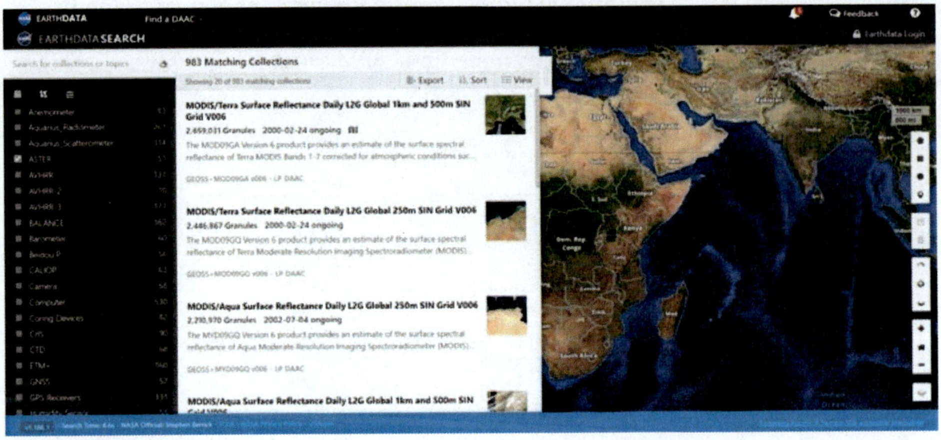

Fig. 3.5 NASA Earthdata Search website

3.6 Planet

PlanetScope, operated by Planet, is a constellation of approximately 130 satellites, able to image the entire land surface of Earth every day (a daily collection capacity of 200 million km^2/day). PlanetScope images are approximately 3 m per pixel resolution. It is available for download at https://www.planet.com/explorer/ (Fig. 3.6) after registration from the following URL: https://www.planet.com/markets/education-and-research/#apply-now.

Fig. 3.6 Planet explorer

3.7 Tellus Traveler

It belongs to the Japanese company and can be accessed from URL https://www.tellusxdp.com/traveler/ (Fig. 3.7). However, you need to register and log in first. There are many physical datasets available for free such as sea surface temperature.

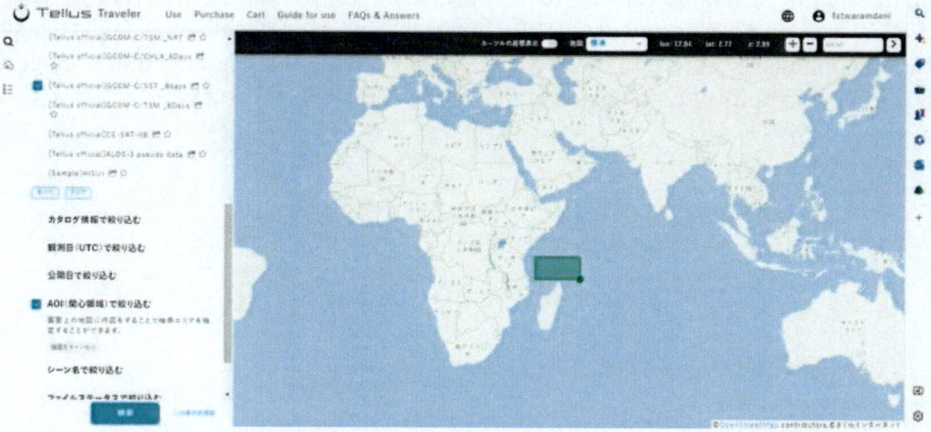

Fig. 3.7 Tellus traveler

3.8 Copernicus Data Space Ecosystem

It is an updated version of Copernicus Open Access Hub. It is an open ecosystem that provides free instant access to a wide range of data and services from the Copernicus Sentinel missions and more on our planet's land, oceans, and atmosphere. The URL is https://dataspace.copernicus.eu/ (Fig. 3.8), and you need to register and log in before downloading the data.

Fig. 3.8 Copernicus Data Space Ecosystem website

QGIS, a powerful geospatial software, provides access to a diverse array of Earth remote sensing satellite imageries, enabling users to harness a multitude of sources for their geospatial analysis and mapping needs. The examples mentioned previously, such as Landsat, Sentinel-2, and MODIS, are just the tip of the iceberg when it comes to the extensive range of imagery sources available in QGIS. Additionally, the software facilitates the exploration and utilization of other prominent sources such as SPOT, IKONOS, QuickBird, GeoEye, and numerous others, further expanding the wealth of remote sensing data at users' fingertips. With QGIS serving as a gateway to such a vast array of satellite imagery sources, researchers, analysts, and mapping enthusiasts are empowered to extract valuable insights, monitor changes over time, and unravel complex spatial patterns across Earth's surface.

Chapter 4
Data Preprocessing

Raster data preprocessing is an indispensable and fundamental stage in the realm of GIS analysis. This pivotal process encompasses a multitude of intricate tasks that are vital for ensuring accurate and reliable results. These tasks, which play a pivotal role in refining the quality and usability of raster data, include but are not limited to meticulous data cleaning to eliminate errors and artifacts, resampling to harmonize varying spatial resolutions, mosaicking to seamlessly merge multiple raster datasets into a cohesive whole, reprojecting to align disparate datasets onto a common coordinate system, and clipping to extract specific regions of interest. To accomplish these complex operations with utmost efficiency and precision, one can harness the power of Quantum Geographic Information System (QGIS), a cutting-edge and open-source GIS software that boasts an extensive suite of tools tailored specifically for raster data preprocessing. With its user-friendly interface and comprehensive functionality, QGIS empowers users to effortlessly navigate through each step of the preprocessing workflow, providing an unparalleled level of control and flexibility. To elucidate the practical implementation of raster data preprocessing using QGIS, we shall delve into a compelling study case revolving around the manipulation of a digital elevation model (DEM). Through this illustrative step-by-step guide, you will gain invaluable insights into the meticulous procedures involved in harnessing the potential of QGIS to preprocess raster data, thereby setting the stage for enhanced GIS analyses and unlocking the true potential of geospatial information.

4.1 Download the Raster DEM Data

To install the OpenTopography DEM plugin, click on Plugin > Manage and Install Plugin (Fig. 4.1).

F. Ramdani, *Exploring the Earth with QGIS*, Springer Remote Sensing/Photogrammetry, https://doi.org/10.1007/978-3-031-46042-5_4

Fig. 4.1 Open plugin manager

Then type "Opentopo," click on "OpenTopography DEM Downloader," then click "Install Plugin (Fig. 4.2)."

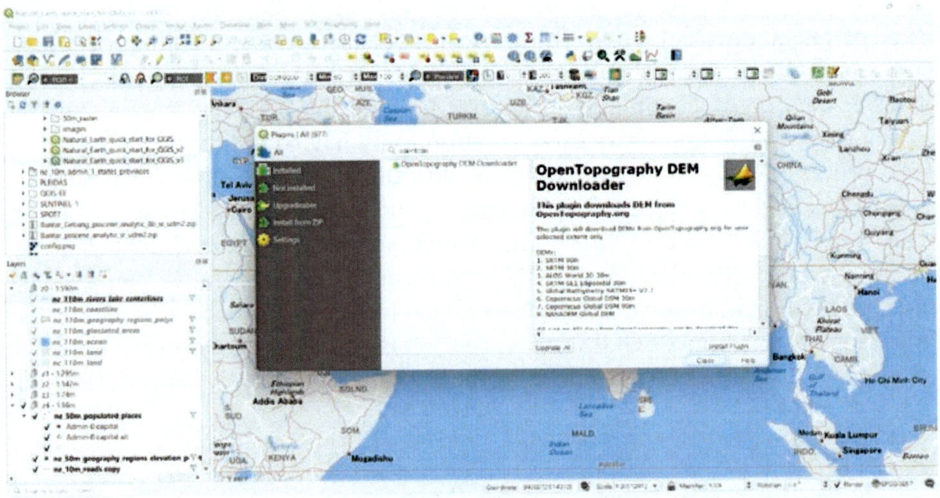

Fig. 4.2 Install OpenTopography DEM plugin

Before you can use the plugin, you need to register and generate a new API key. Register here https://portal.opentopography.org/newUser then after login go to https://portal.opentopography.org/myopentopo

Scroll down and click on "myOpentopo Authorizations and API Key." Click on "Request API" then copy your key (Fig. 4.3). In the new website interface, you may click on "My Open Topo" and then click on "Get an API Key" then copy your key.

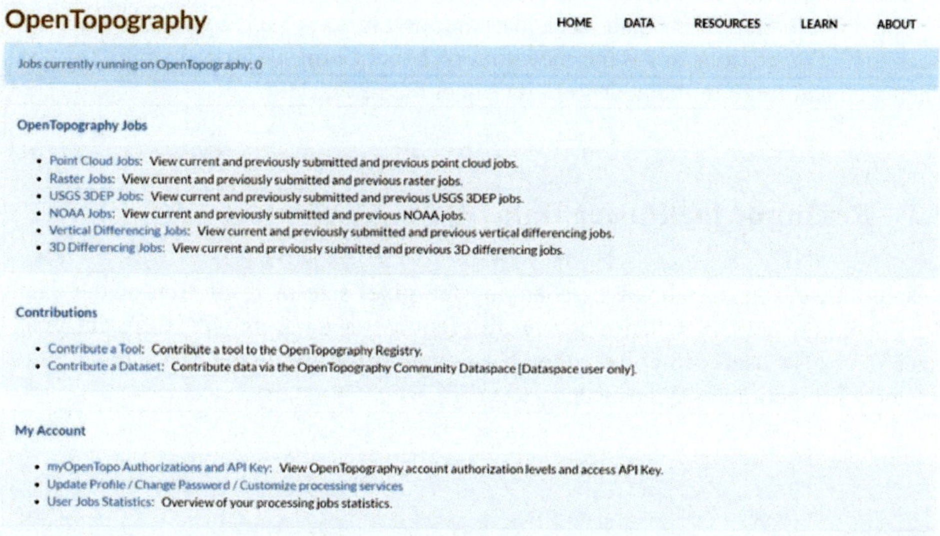

Fig. 4.3 Requesting API

Go back to QGIS, open the OpenTopo plugin by clicking the icon .

Select the DEM data to download, define the extent to download (Use Map Canvas Extent), and paste your API key in the "Enter your API key…" then click Run (Fig. 4.4). The maximum area for download DEM data using OpenTopo plugin is 4,050,000 km^2, so make sure your study area is less then 4,000,000 km^2.

In this case, we zoom in around Mount Fuji, Japan.

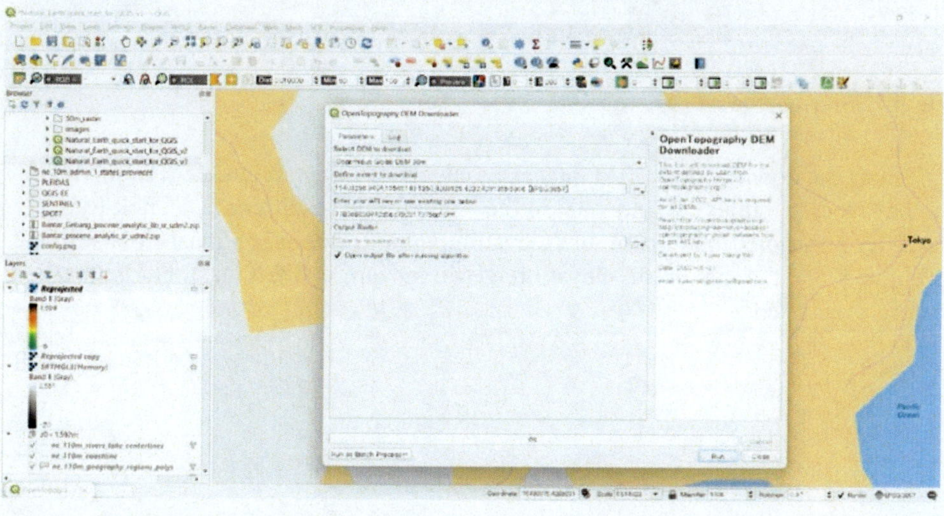

Fig. 4.4 Open Topography DEM downloader

4.2 Check Data Quality

It is essential to check the quality of the data before performing any analysis. Check for data completeness, pixel size, resolution, elevation values, etc.

The downloaded DEM data has a maximum elevation of 2552 m above sea level (masl). It is suspicious since the elevation of Mount Fuji is more than 3000 masl. Then we need to go to the next step, reproject the raster data.

4.3 Resample the Raster Data

Resampling raster data involves changing the pixel size or resolution of the data. This task is necessary when merging different raster datasets with different pixel sizes. It is also useful to correct the DEM data. To resample the raster data, go to Raster > Projections > Warp (Reproject) (Fig. 4.5).

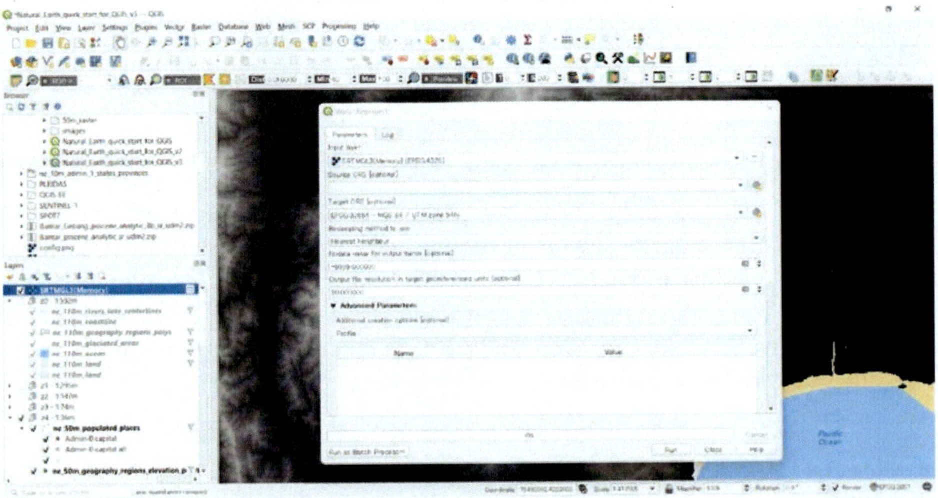

Fig. 4.5 Warp (reproject)

Select the downloaded DEM data as input, change the target CRS, input "-9999" in the "Nodata values…" and "30" in the "Output file resolution…" Click Run.

After resampled, the maximum elevation value is now 3558 masl. It is still less accurate since the maximum elevation of Mount Fuji is 3776 masl, but better than before resampled.

4.4 Clip the Raster Data

Clipping raster data is essential when you need to extract a specific area or region of interest from a larger dataset. To clip the raster data, go to Raster > Extraction > Clip Raster by Mask Layer.

We will use polygon boundary from Natural Earth package data (available at https://www.naturalearthdata.com/downloads/). Open layer "ne_10m_admin_1_states_provinces" and select only "Yamanashi" Province as shown in the Fig. 4.6.

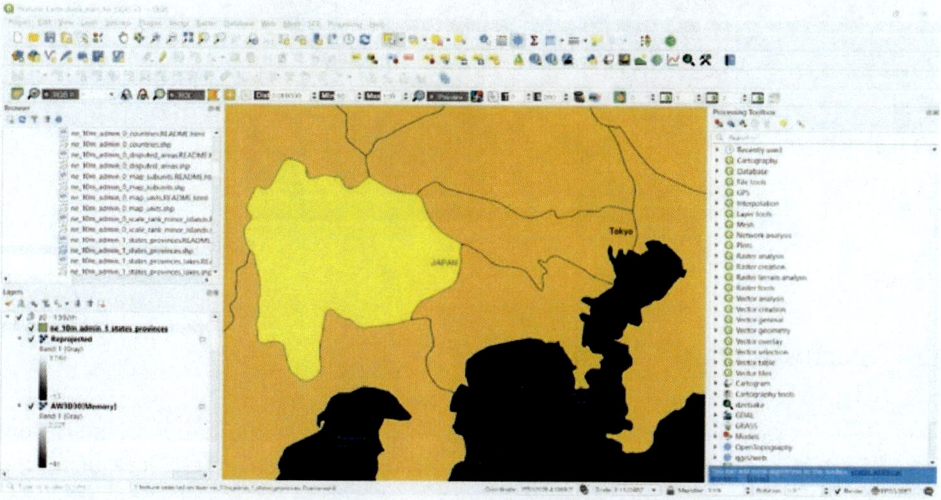

Fig. 4.6 Selecting study area

Then go to Raster > Extraction > Clip Raster by Mask Layer, and check on "Selected features only." Select the "Reprojected" data as Input Layer then click Run (Fig. 4.7).

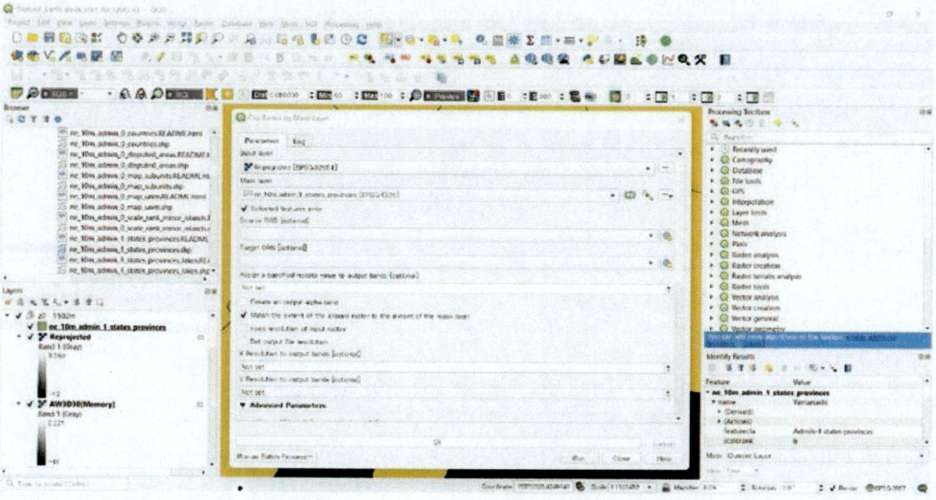

Fig. 4.7 Clip raster by mask layer

The result will appear as shown in Fig. 4.8.

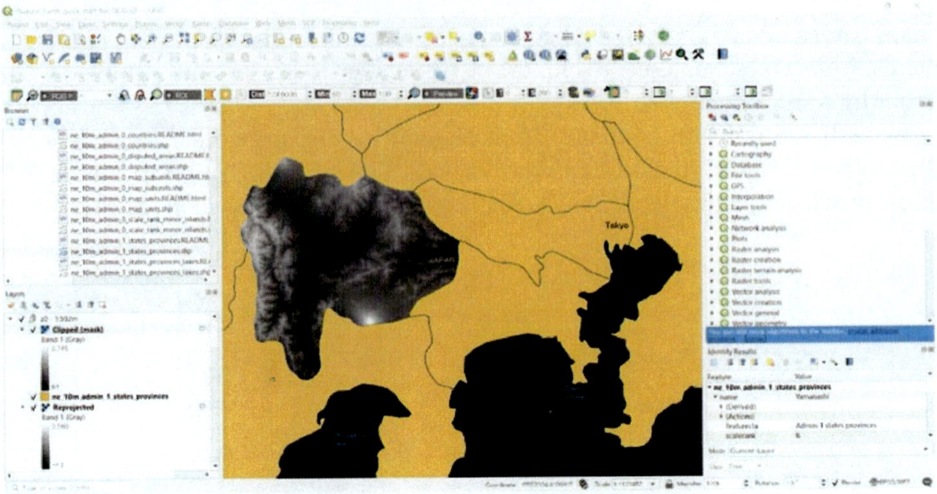

Fig. 4.8 Clipped result

4.5 Perform Data Cleaning

Data cleaning involves removing or repairing any errors or anomalies in the raster data. Use the various tools provided in the QGIS toolbox, such as Fill NoData, to fill in missing data or gaps in the data.

Activate the Processing Toolbox from Processing > Toolbox. Type "Fill" and select "Fill NoData cells." Select "Clipped (mask)" as Raster Input and type "1" in the Fill value, then click Run (Fig. 4.9).

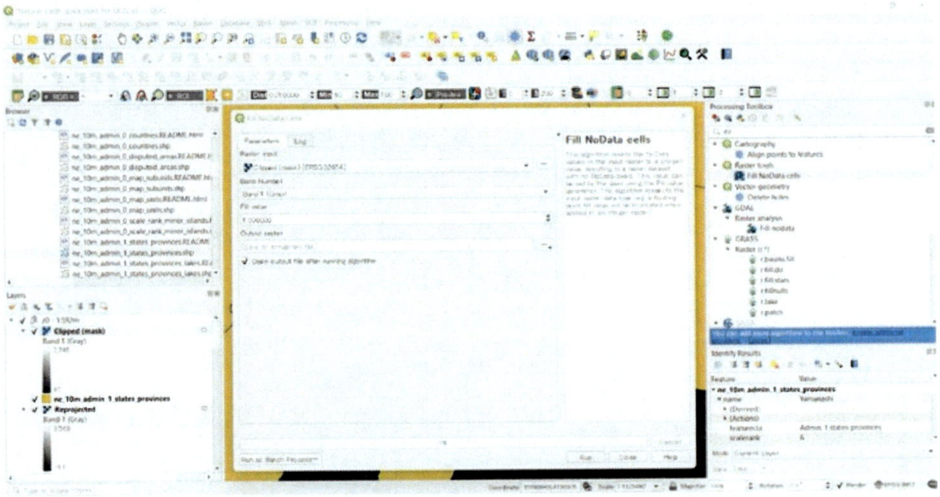

Fig. 4.9 Fill NoData cells

4.6 Apply Color and Styling

Finally, apply color and styling to the raster data to enhance its visualization. Go to Layer > Properties > Symbology to adjust the color and style of the raster data.

Double-click on the "Output raster" layer, select Symbology then change the "Render type" to "Hillshade," and Blending mode to "Multiply," and click OK (Fig. 4.10).

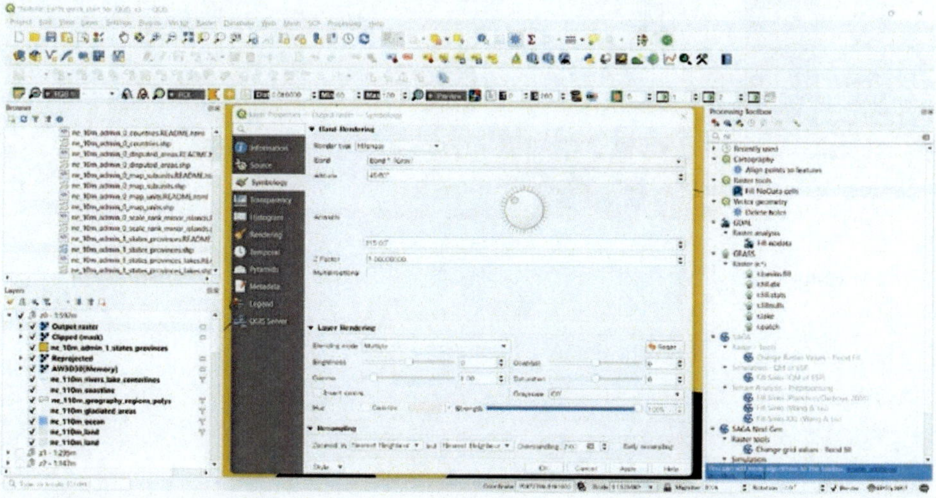

Fig. 4.10 Layer properties

Then right-click on the "Output raster" layer, and select "Duplicate Layer."

Double-click on the "Output raster copy," and change the Render type to "Singleband pseudocolor." On the "Color ramp," click on the small drop-down arrow and select "Create new color ramp" > "Catalog: cpt-city."

Scroll down and select "Topography" > "elevation," then click OK (Fig. 4.11).

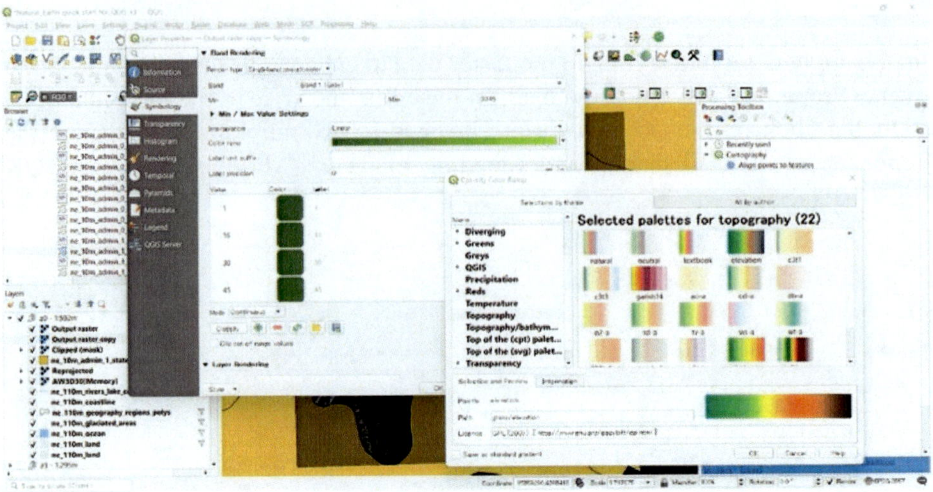

Fig. 4.11 Color styling

The final result will be shown as Fig. 4.12.

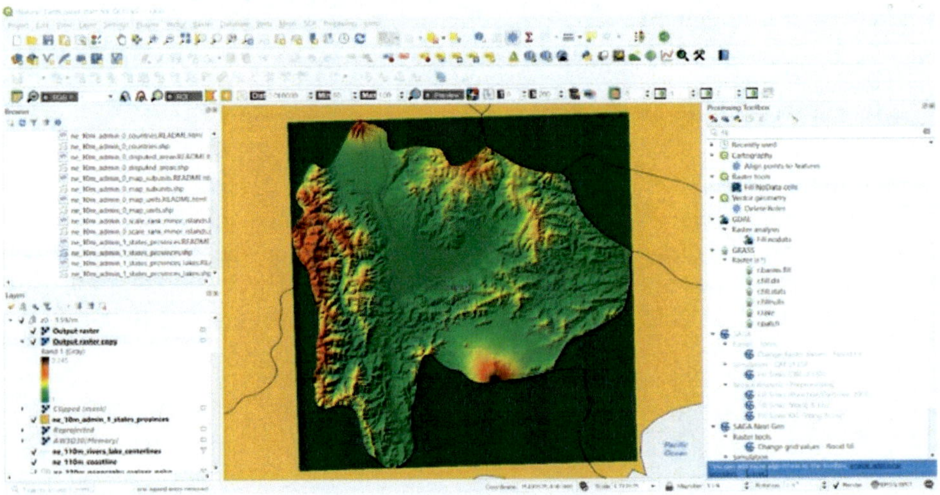

Fig. 4.12 Result of color styling

This dataset is still temporary, you can save the DEM data (Fig. 4.13) in your folder by right clicking on the "Output raster" layer, and select Export > Save as.

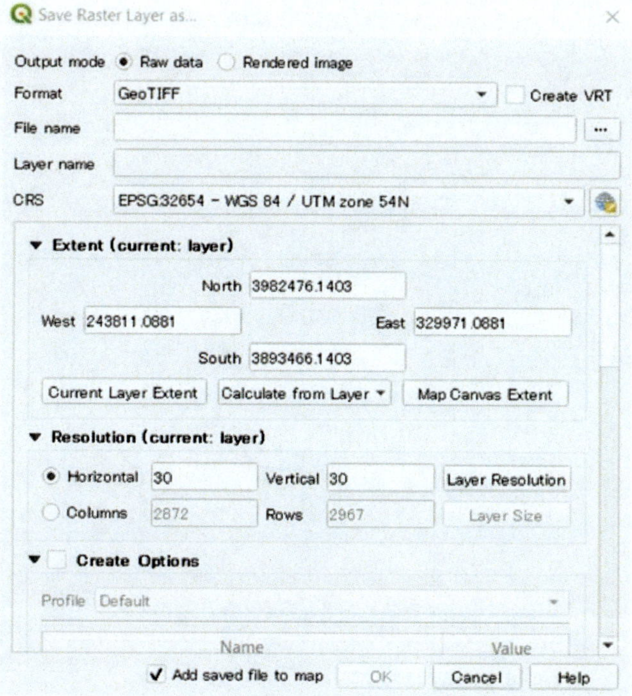

Fig. 4.13 Export DEM data

Chapter 5
Data Management

QGIS, a robust and feature-rich open-source geographic information system (GIS) software, provides an exceptional platform for effectively managing and analyzing vast quantities of remote sensing satellite imagery data. With its comprehensive set of tools and capabilities, QGIS empowers users to seamlessly handle the complex and intricate tasks associated with remote sensing data. By leveraging QGIS, professionals and researchers can efficiently organize, store, retrieve, and manipulate satellite imagery datasets, allowing them to delve deep into the wealth of information embedded within these geospatial resources. Moreover, QGIS facilitates the integration of diverse data formats and sources, enabling users to effortlessly harmonize and consolidate information from multiple sensors and platforms.

Through its intuitive user interface and extensive library of plugins, QGIS further enhances the data management process by providing a user-friendly environment where individuals can easily visualize, explore, and navigate through their satellite imagery collections. Whether it's conducting sophisticated image analysis, performing spatial queries, or generating accurate thematic maps, QGIS stands as a versatile and indispensable tool in the realm of remote sensing, empowering users to unlock valuable insights and make informed decisions based on the wealth of satellite imagery data at their disposal.

5.1 Importing Data

QGIS allows users to import various types of satellite imagery data, such as GeoTIFF, NetCDF, and HDF5, as well as other data formats like shapefiles and GPS data. In this case, we will download and import satellite data into QGIS.

F. Ramdani, *Exploring the Earth with QGIS*, Springer Remote Sensing/Photogrammetry, https://doi.org/10.1007/978-3-031-46042-5_5

Using the administrative boundary of Yamanashi, go to Natural_Earth_quick_start > packages > Natural_Earth_quick_start >10m_cultural > ne_10m_admin_1_states_provinces.shp. We will download the Sentinel-2 dataset from Copernicus Data Space Ecosystem (available at https://dataspace.copernicus.eu/).

Click on Yamanashi using the "Select features" icon then go to Layer > Save as and then save as GeoJSON (Fig. 5.1).

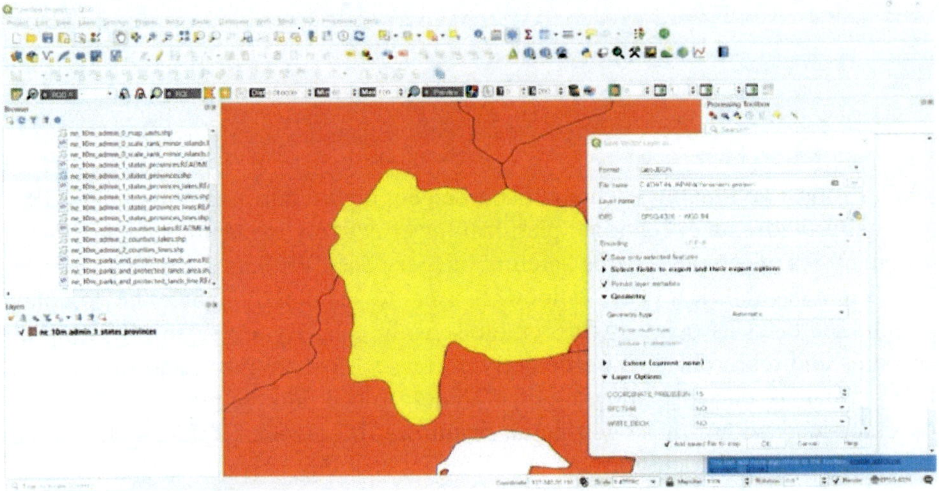

Fig. 5.1 Select and save features

Open your Internet browser then open the Copernicus Data Space Ecosystem website. Login and type "Yamanashi" in the search window. Click on the "Create an area of interest" icon then upload your "Yamanashi" GeoJSON file (Fig. 5.2).

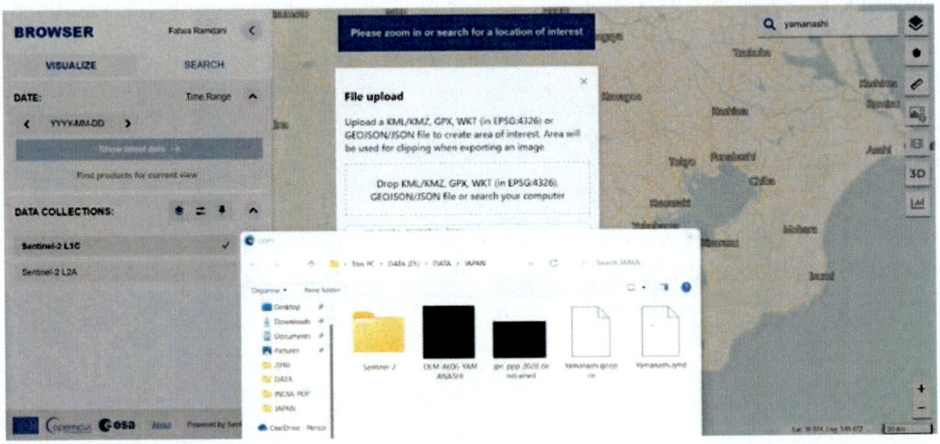

Fig. 5.2 Uploading area of interest

Click on "SEARCH" then check on Sentinel-2 data, adjust the cloud cover percentage to less than 30%, then click "Search" (Fig. 5.3).

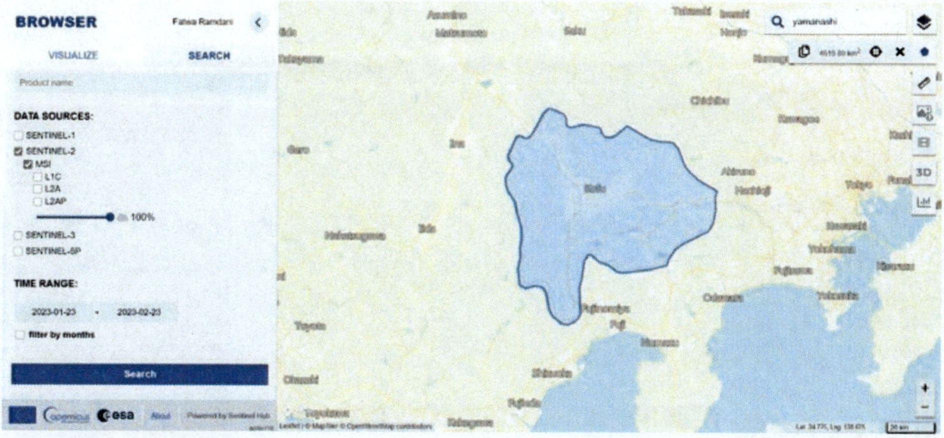

Fig. 5.3 Searching datasets

Choose any layer and click "Visualize" on the left panel. You can change the visualization to "Highlight Optimized Natural Color." Then click on the "Download image"  icon to download (Fig. 5.4).

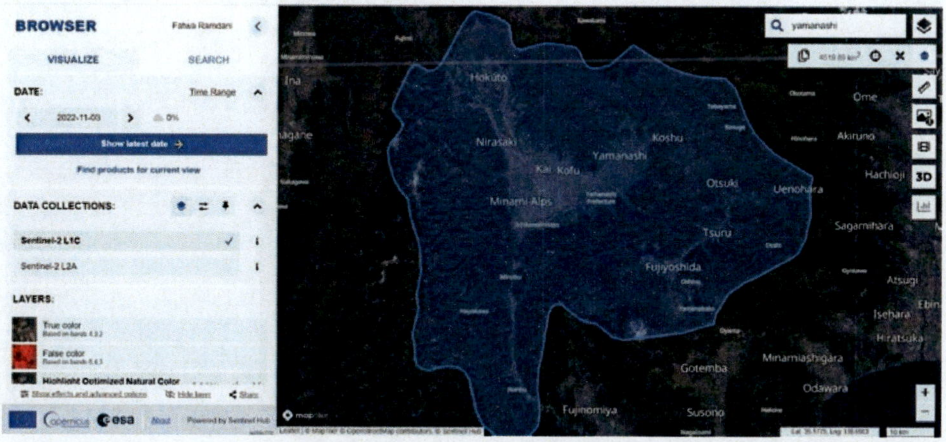

Fig. 5.4 Changing the visualization and download image

Click on "Analytical," change the "Image format" to TIFF (8-bit), "Image resolution" to MEDIUM, and "Coordinate systems" to WGS 84 (EPSG:4326). Check on "B08" and "B11," scroll down and click Download (Fig. 5.5).

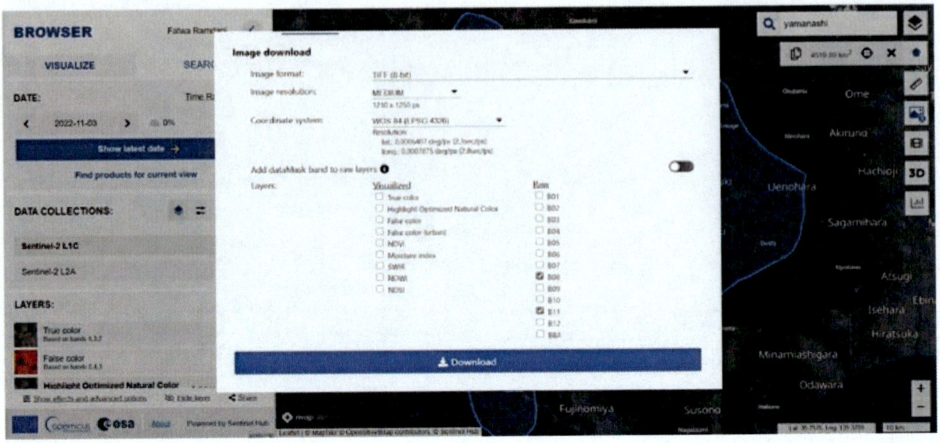

Fig. 5.5 Changing the image format

Go back to your QGIS, extract first then open the downloaded files (Fig. 5.6).

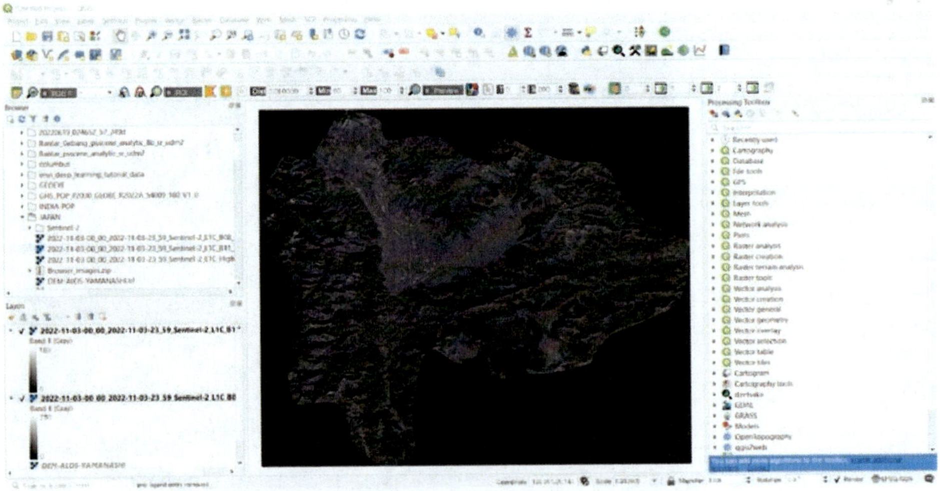

Fig. 5.6 Open the downloaded data in QGIS

5.2 Classification

QGIS has tools for the classification of satellite imagery data. Users can use these tools to identify and classify different features in the imagery, such as water bodies, urban areas, and vegetation. For topography mapping, the user can define the elevation into different classes such as low, medium, and high elevation. While for urban mapping, users can define urban density into some classes such as low, medium, and high urban density.

In this case, we will classify the urbanized area using the Normalized Difference Built-Up Index (NDBI). NDBI is a remote sensing index that is commonly used to identify built-up areas in urban regions. It is based on the principle that built-up areas have a different spectral signature compared to non-built-up areas, such as vegetation or bare soil.

NDBI is calculated using the following formula:

$$\text{NDBI} = (\text{NIR} - \text{SWIR}) / (\text{NIR} + \text{SWIR})$$

Where NIR is the reflectance in the Near Infrared band, and SWIR is the reflectance in the Shortwave Infrared band.

In this formula, the NIR band is used as a proxy for the non-built-up features, such as vegetation or water, which have a high reflectance in this band. The SWIR band, on the other hand, is used as a proxy for built-up features, such as roads, buildings, and other impervious surfaces, which have a low reflectance in this band.

The output of the NDBI calculation ranges from −1 to 1, with positive values indicating built-up areas and negative values indicating non-built-up areas. A threshold value can be set to classify pixels as either built-up or non-built-up.

NDBI is widely used for urban mapping and land-use classification, as well as for monitoring urban growth and land-use changes over time. It can also be used in combination with other remote sensing indices to improve the accuracy of urban mapping and land-use classification.

We will calculate using only B08 (NIR) and B11 (SWIR) of Sentinel-2. Go to Raster > Raster calculator, then type the following expression:

```
("2022-11-03-00_00_2022-11-03-23_59_Sentinel-2_L1C_B11_(Raw)@1" -
"2022-11-03-00_00_2022-11-03-23_59_Sentinel-2_L1C_B08_(Raw)@1"  )
/ ( "2022-11-03-00_00_2022-11-03-23_59_Sentinel-2_L1C_B11_(Raw)
@1"   +   "2022-11-03-00_00_2022-11-03-23_59_Sentinel-2_L1C_B08_
(Raw)@1" )
```

The file name can be different, but the principle is the formula. Give it the name and save it into your folder, then click OK (Fig. 5.7).

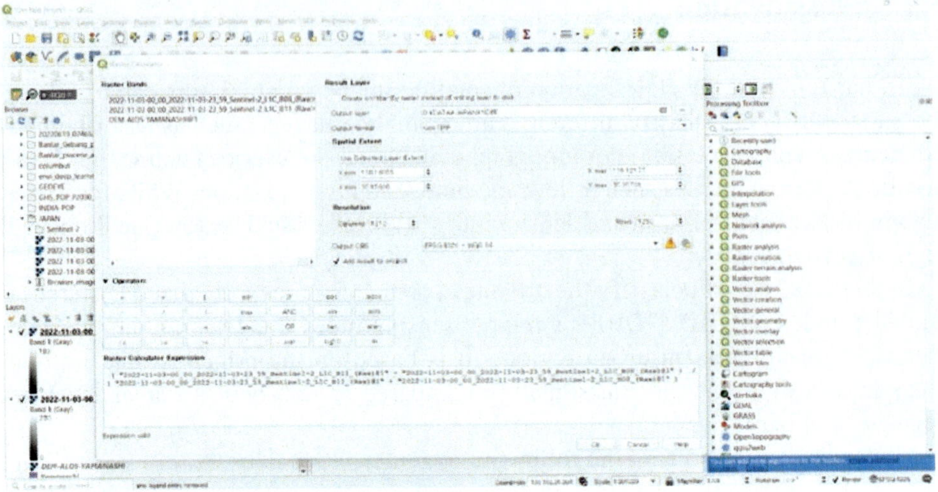

Fig. 5.7 Raster calculator

The output will display values ranging from −1 to 1. A higher value indicates a greater degree of urbanization, while a lower value indicates a lesser degree of urbanization.

Then open Processing Toolbox, type “reclass” and double-click on “Reclassify by table.” On the open window, click on the icon ••• next to “Reclassification table.” Add five rows then input the values as follows. Click OK then Run (Fig. 5.8).

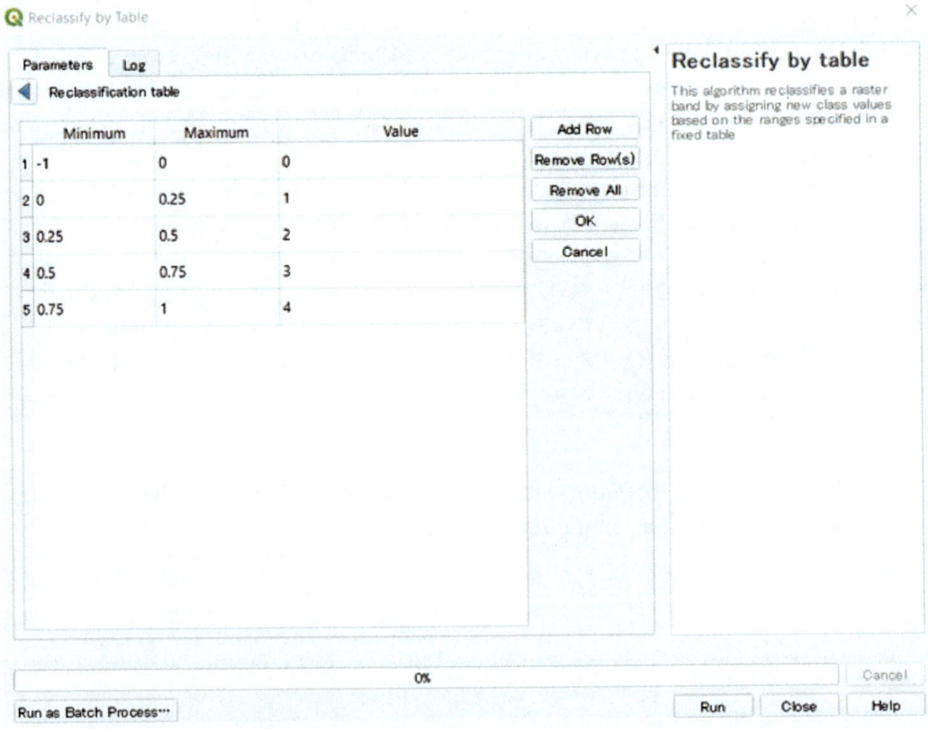

Fig. 5.8 Reclassify by using table

You can change the visualization by double-clicking on the "Reclassified raster" and then changing the Render type to "Palette/Unique values." Change the opacity level to 0% except for value 1 (Fig. 5.9).

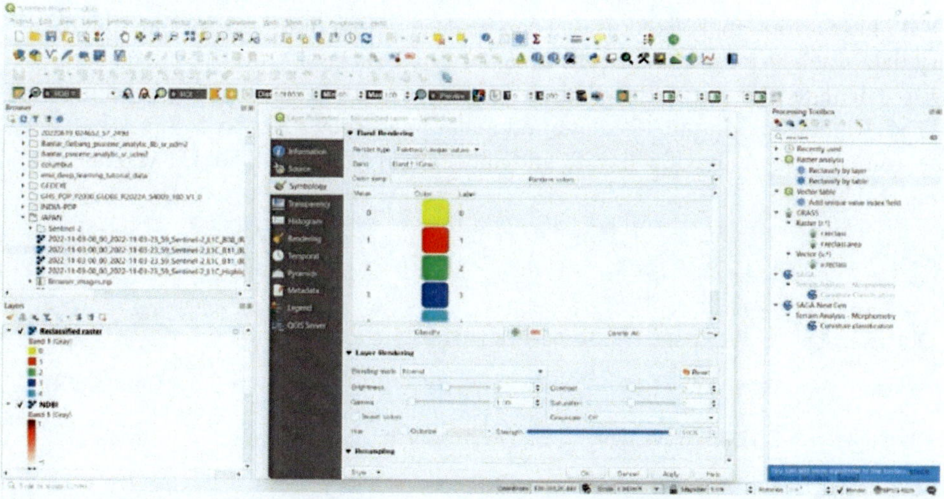

Fig. 5.9 Adjusting the visualization

The result will be shown as Fig. 5.10.

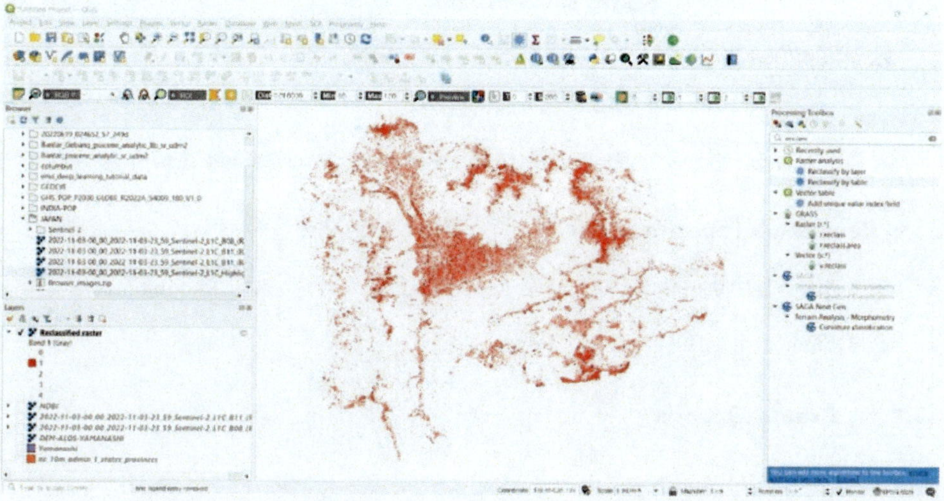

Fig. 5.10 Result of only urbanized area

5.3 Analysis

5.3.1 Satellite Image Analysis

QGIS has various analysis tools that can be used with remote sensing satellite imagery data, such as calculation of the area. After we reclassify the urbanized area, we will calculate the area. But before that, we need to convert from degree to UTM. Right-click on the "Reclassified raster" and select Export > Save as. Select the folder, give it the file name, and then change the CRS to UTM zone 54N (Fig. 5.11).

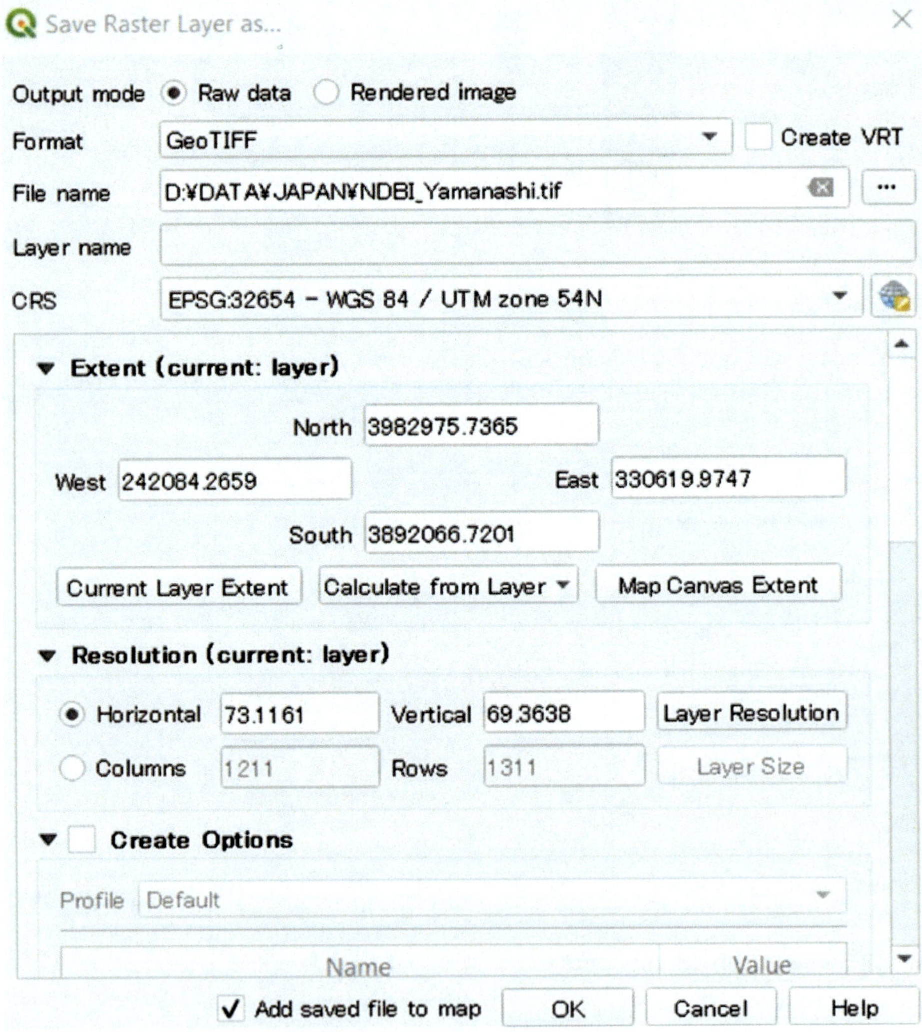

Fig. 5.11 Changing raster projection

Go back to Processing Toolbox then type "value repo," double-click on "Raster layer unique value report" then click Run (Fig. 5.12).

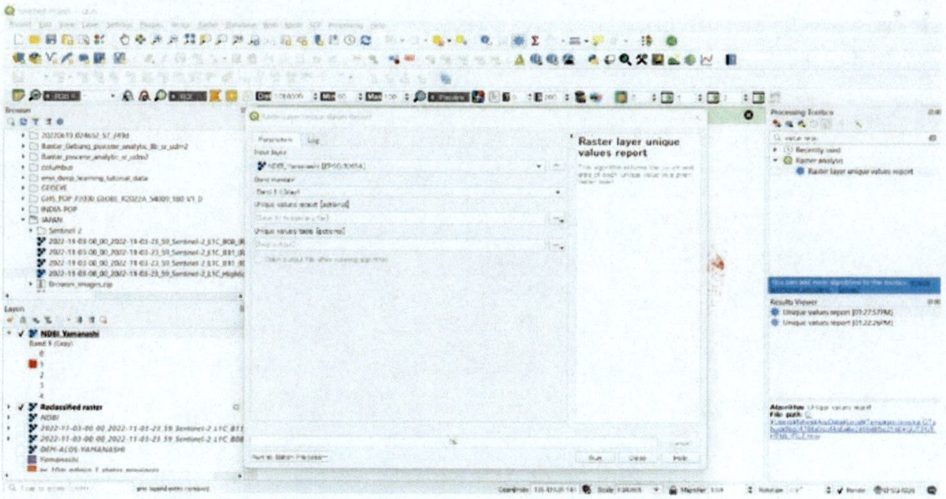

Fig. 5.12 Raster layer unique values report

Click on the "Result viewer"

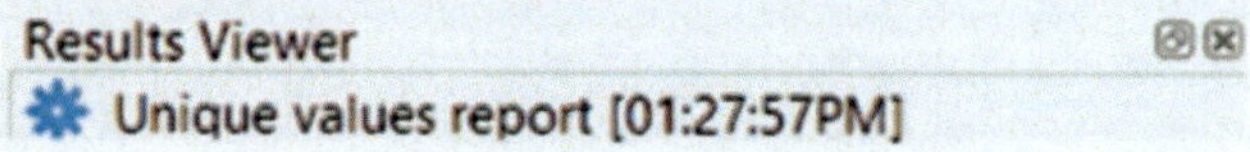

.

The result will be shown in a browser like this.

Analyzed file: D:\DATA\JAPAN\NDBI_Yamanashi.tif (band 1)

Extent: 242084.2658999999985099,3892066.7201000000350177 : 330619.9747000000206754,3982975.7365000001154840

Projection: EPSG:32654 - WGS 84 / UTM zone 54N

Width in pixels: 1211 (units per pixel 73.1096)

Height in pixels: 1311 (units per pixel 69.3433)

Total pixel count: 1587621

NODATA pixel count: 695161

Value	Pixel count	Area (m^2)
-1	217	1100115.608267218
0	747314	3788625786.528145
1	139320	706304638.4506395
2	4457	22595462.05551608
3	446	2261067.102705895
4	706	3579177.969754174

You can also save the result in Table, click on the icon next to "Unique values table [optional]" and then save it to Geopackage (Fig. 5.13).

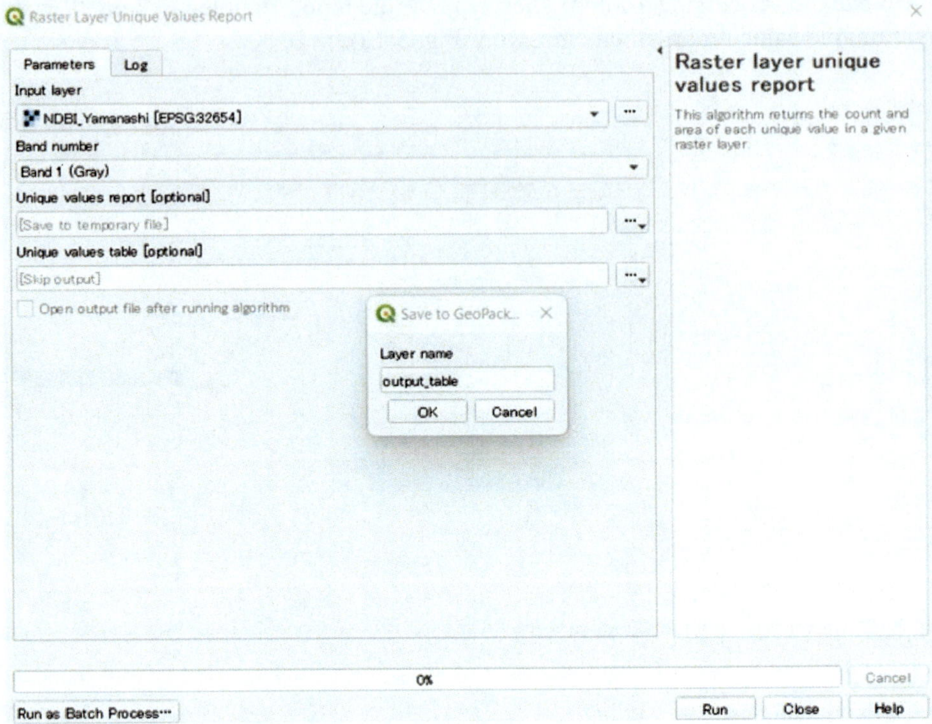

Fig. 5.13 Save as GeoPackage

After that you can see the "output_table" on the left layer panel (Fig. 5.14).

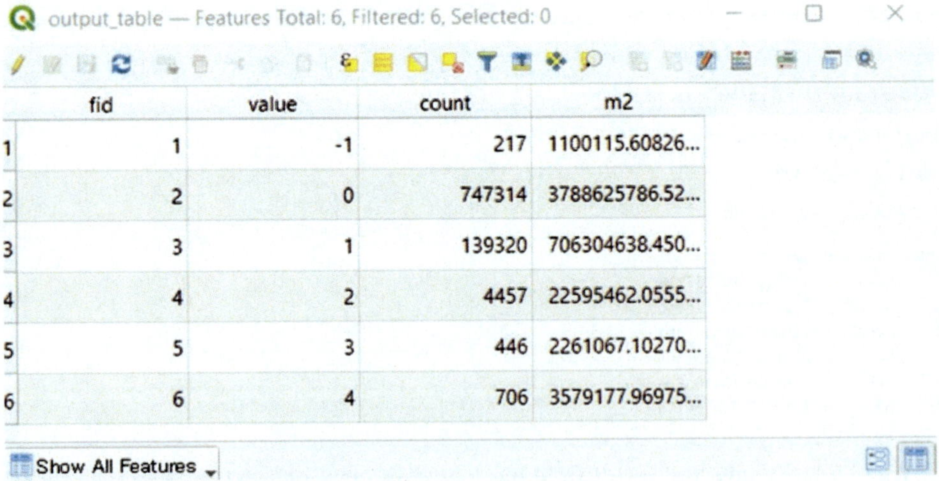

	fid	value	count	m2
1	1	-1	217	1100115.60826...
2	2	0	747314	3788625786.52...
3	3	1	139320	706304638.450...
4	4	2	4457	22595462.0555...
5	5	3	446	2261067.10270...
6	6	4	706	3579177.96975...

Fig. 5.14 Calculation result

To compute the area in square kilometers, access the "Field Calculator" in the Processing Toolbox by double-clicking on it. Enter the desired field name, such as "sqkm," and then input the following expression:

```
round ( "m2" / 1e6, 2)
```

We will convert the unit from square meters to square kilometers and round the result to two decimal places (Fig. 5.15).

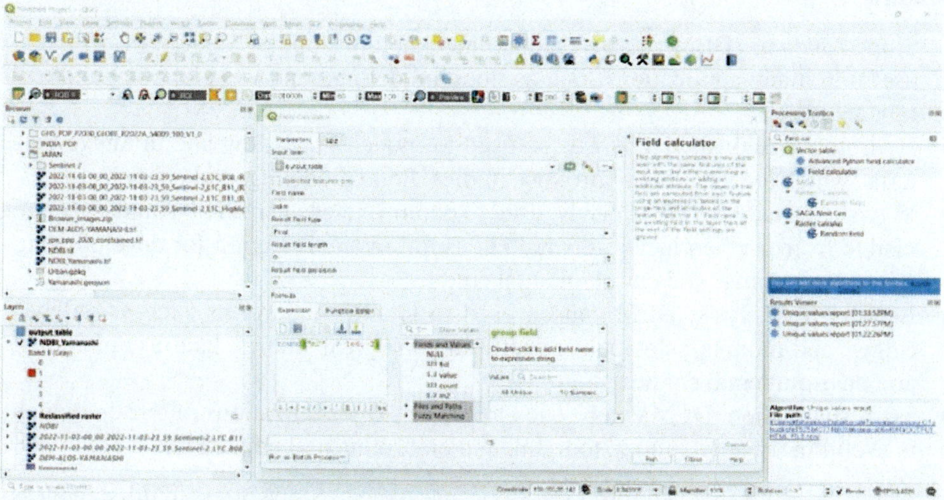

Fig. 5.15 Convert the unit from square meters to square kilometers

Now we get the result of the urbanized area of Yamanashi for the year 2023 is 706.3 km^2 (Fig. 5.16).

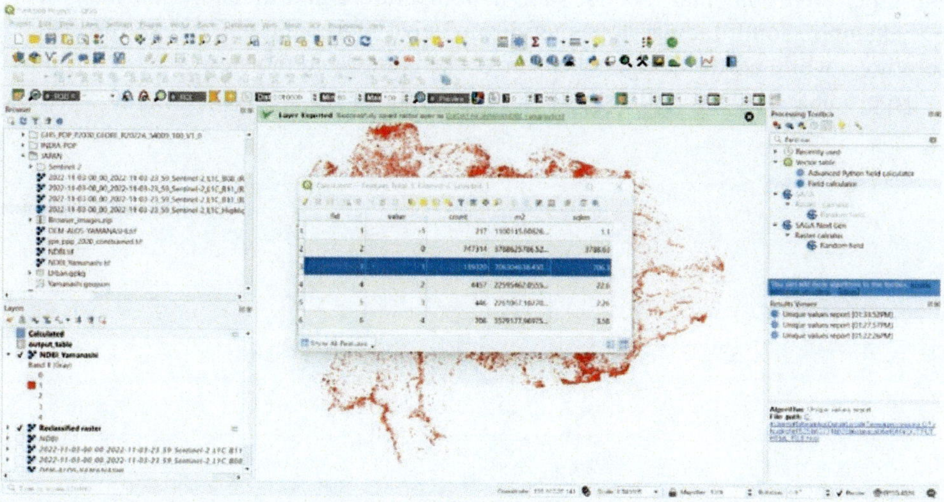

Fig. 5.16 Result in square kilometers

5.3.2 DEM Analysis

DEM stands for Digital Elevation Model, which is a digital representation of the elevation of the Earth's surface. It is typically created by taking elevation measurements of the Earth's surface and organizing them into a regular grid of cells, with each cell containing information about the elevation at that point.

DEM is a powerful tool in GIS analysis, as it allows us to understand and analyze the topography of an area. Some examples of its application using GIS are as follows:

- Slope Analysis: DEM can be used to calculate the slope of the terrain, which is useful in many applications such as flood modeling, land-use planning, and terrain analysis.
- Hydrology Analysis: DEM can be used to model water flow and drainage patterns, which are useful in hydrology studies, flood modeling, and erosion studies.
- Viewshed Analysis: DEM can be used to determine what areas of a landscape are visible from a given location, which is useful in site selection for development, military planning, and environmental studies.
- Landform Analysis: DEM can be used to identify landforms such as valleys, ridges, and mountains, which is useful in geological studies, land-use planning, and environmental studies.
- 3D Visualization: DEM can be used to create 3D visualizations of terrain, which is useful in urban planning, tourism, and environmental education.

Using DEM data we will produce Slope and Aspect. Slope refers to the steepness of the terrain, which is defined as the angle between the surface and a horizontal plane. The slope is usually expressed as a percentage or an angle in degrees, and it can range from 0 (flat) to 90 (vertical). While Aspect refers to the direction that a slope faces, it is measured in degrees clockwise from the north and can range from 0 to 360 degrees. Aspect is important because it affects the distribution of sunlight and other environmental factors, such as temperature and moisture, on the slope. For example, south-facing slopes tend to receive more sunlight and warmer temperatures, while north-facing slopes tend to be cooler and moister.

To produce Slope go to Raster > Analysis > Slope (Fig. 5.17).

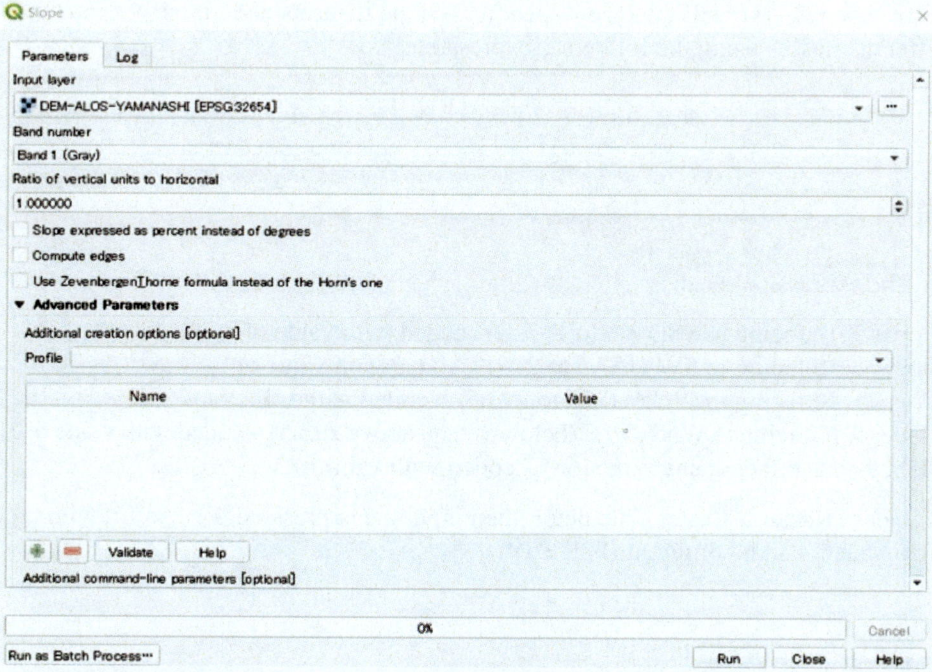

Fig. 5.17 Produce slope

To produce Aspect go to Raster > Analysis > Aspect (Fig. 5.18).

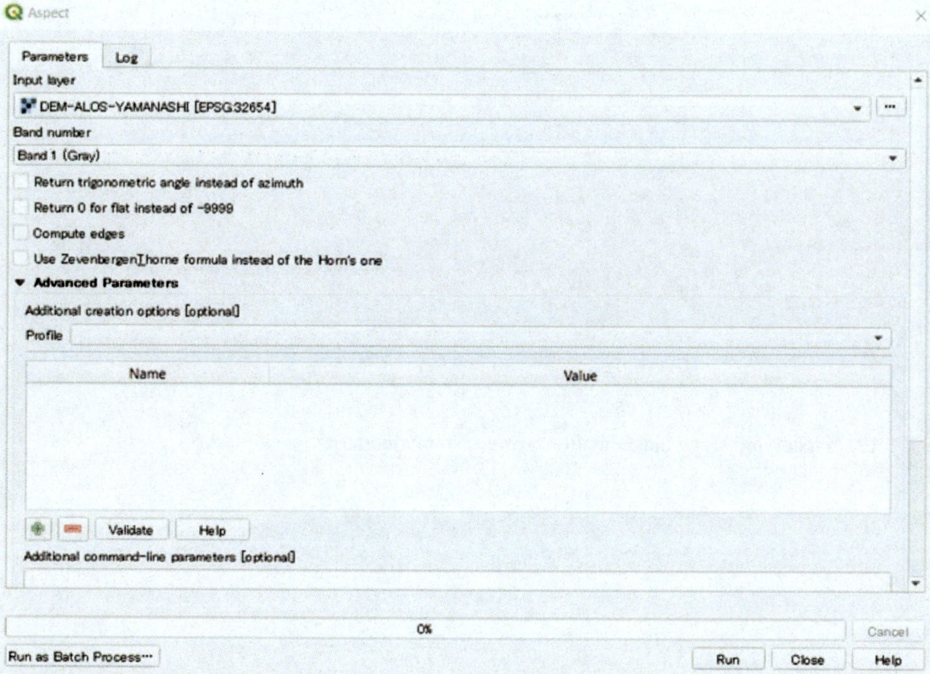

Fig. 5.18 Produce aspect

In this case, we will analyze Aspect and Slope datasets and classify it into five different classes using the following expression:

```
( ( ("Aspect@1" < 90) OR ("Aspect@1" > 292.5) ) * 5 + ( ("Aspect@1"
>= 90) AND ("Aspect@1" <157.5) ) * 3 + ( ("Aspect@1" >= 157.5) AND
("Aspect@1" < 225) ) * 2 + ( ("Aspect@1" >= 225) AND ("Aspect@1"
<= 292.5) ) * 4 ) * ("Slope@1" > 10) * 1
```

The expression means:

Slopes > 10 facing to SSE (90 to 157.5) – coded with value of 3
Slopes > 10 facing to SW (157.5 to 225) – coded with value of 2
Slopes > 10 facing to WNW (225 to 292.5) – coded with value of 4
Slopes > 10 facing to WNW to E (below 90 and above 292.5) – coded with value of 5
Slopes <= 10 (facing any direction) – coded with value of 1

Got to Raster > Raster Calculator, then input the expression. Be careful with the file name, it can be different (Fig. 5.19).

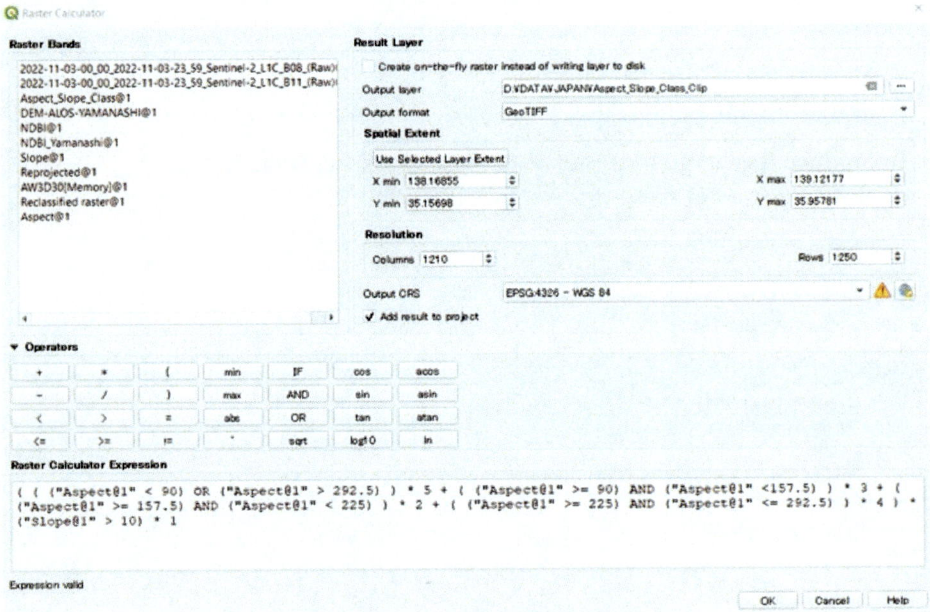

Fig. 5.19 Aspect and slope classification using raster calculator

The result is as shown in Fig. 5.20.

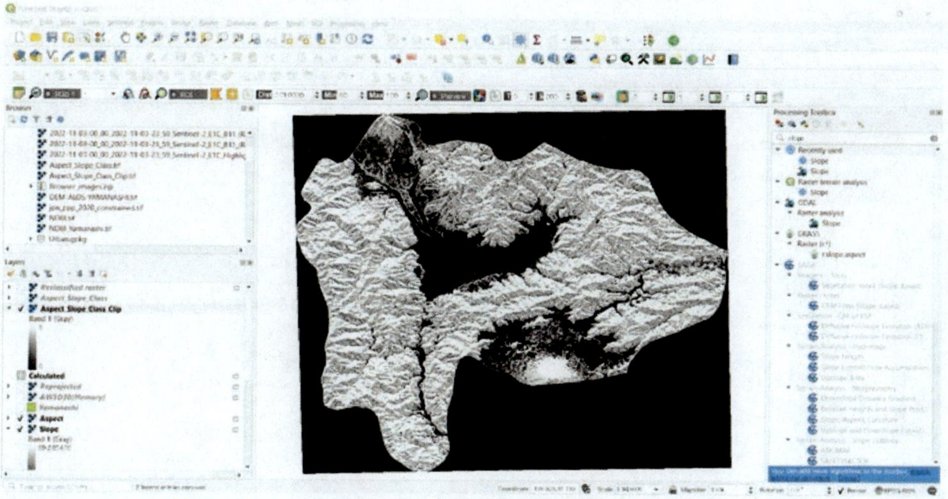

Fig. 5.20 Result of aspect and slope classification

5.4 Visualization

QGIS has a range of tools for visualizing remote sensing satellite imagery data, including color balancing, contrast enhancement, and image blending. These tools allow users to create visually appealing maps and images for presentations or publications.

Using the DEM of Yamanashi Prefecture, we can blend the DEM Image with the classified Aspect and Slope. Arrange the layer so the classified Aspect and Slope layer is located above the DEM layer. Double-click on the DEM layer and change the Render type to "Singleband pseudocolor." Click on the small arrow next to "Color ramp" then select Create new color ramp > Catalog: cpt-city. Then scroll down, click on Topography > elevation, then click OK-Apply (Fig. 5.21).

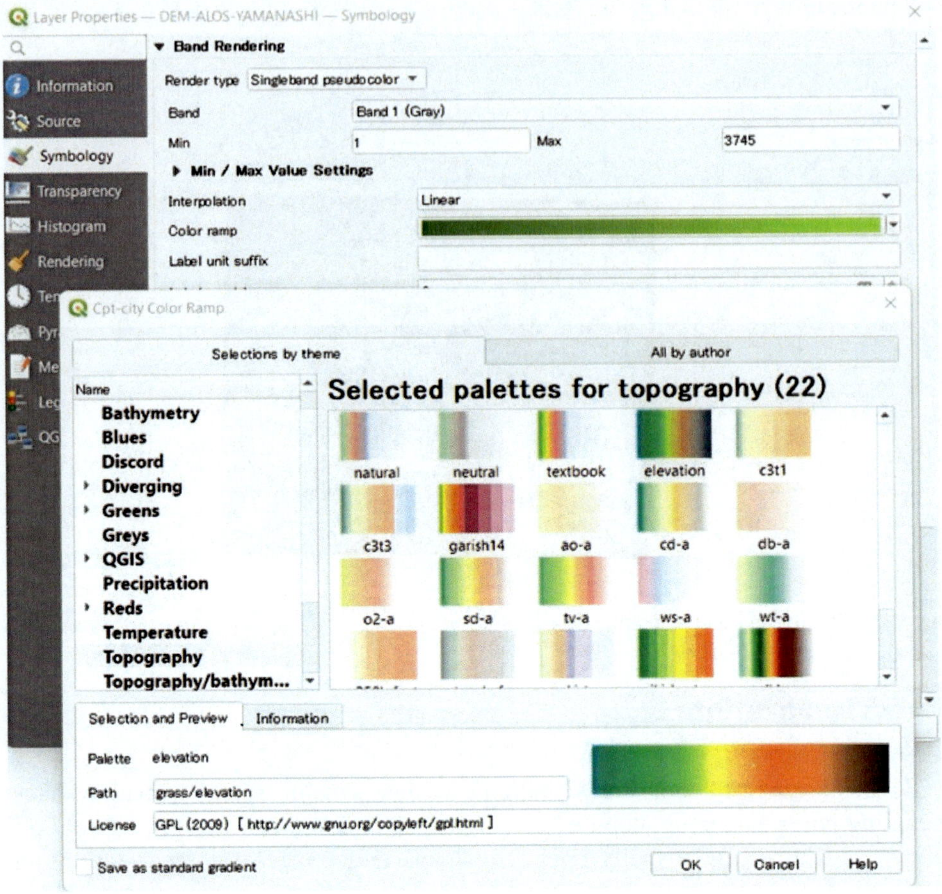

Fig. 5.21 Color adjustment

Change both layers' blending mode to "Multiply," go to Layer Rendering > Blending mode > Multiply and then click OK. The final result will be as shown in Fig. 5.22.

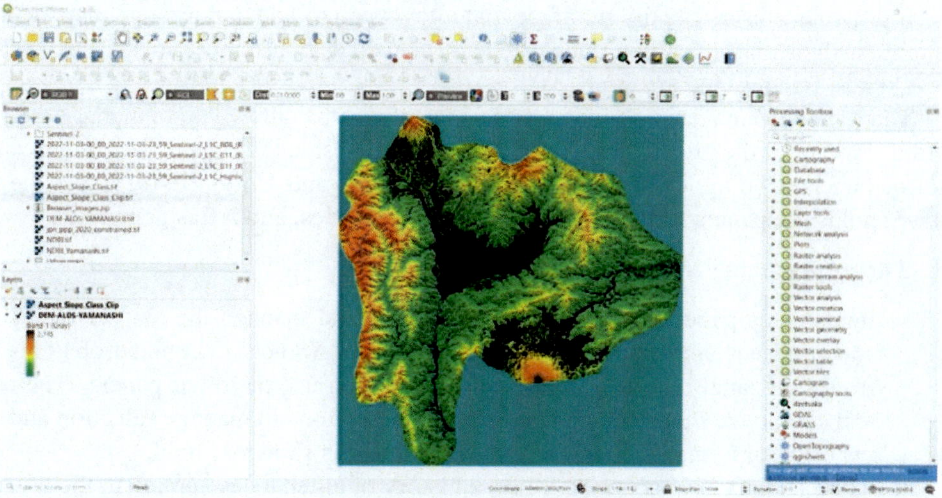

Fig. 5.22 Final result of aspect and slope after color adjustment

5.4.1 Layout Manager

One of the key features of QGIS is its ability to create and manage layouts, which are essentially map compositions that can be printed or exported to various formats. To create a layout in QGIS, go to Project > Layout Manager > Create > OK (Fig. 5.23). The Layout Manager is a tool that allows you to create, edit, and manage layouts within QGIS.

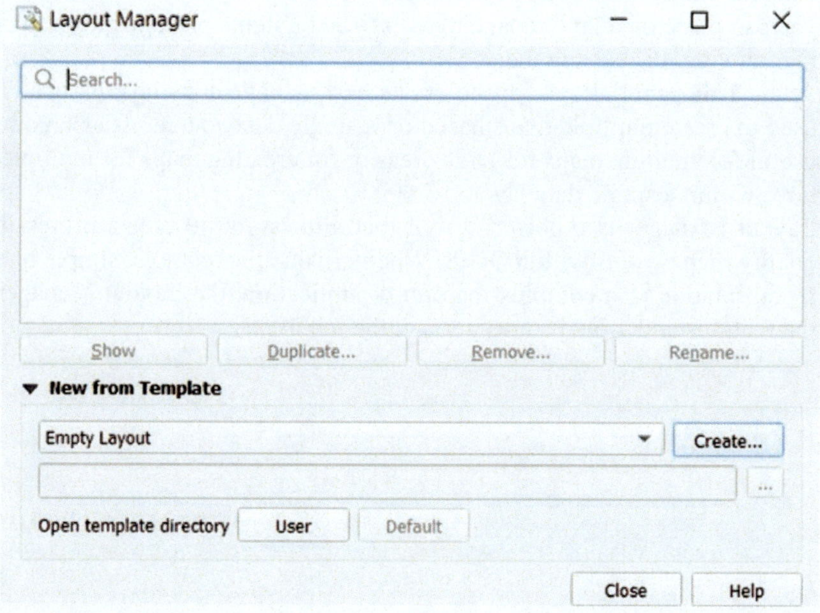

Fig. 5.23 Create layout

It provides a user-friendly interface that allows you to add and arrange map elements, such as maps, legends, and scale bars, as well as non-map elements, such as text boxes, images, and arrows.

The Layout Manager is divided into several panels that allow you to manage different aspects of the layout (Fig. 5.24). On the right beside the canvas, you find two set of panels. The upper one holds the panels **Items** and **Undo History,** and the lower holds the panels **Layout, Item properties, Guides,** and **Atlas** generation.

1. **The upper panels:**
   - Items: This panel allows you to add, delete, and manage the map elements and non-map elements that are part of the layout. You can drag and drop items from this panel onto the layout canvas and then use the other panels to edit and customize them. This panel offers some options to manage selection and visibility of items, and note that a selected row is shown as bold.
   - Undo History: This panel displays a history of all changes applied to the layout. With a mouse click, it is possible to undo and redo layout steps back and forth to a certain status.
2. **The lower panels:**
   - Layout: This panel allows you to define the global settings of your print layout and manage the properties of the layout itself, such as its size, orientation, and background color. You can also add guides to help you align and position map elements on the canvas.
   - Item Properties: This panel displays the properties for the selected item. Click the Select/Move item icon to select an item (e.g., legend, scale bar or label) on the canvas. Then click the Item Properties panel and customize the settings for the selected item.
   - Guides: This panel allows you to adjust vertical or horizontal line references, you can place on a layout page to assist you on items placement, when creating, moving, or resizing them.
   - Atlas: This panel allows you to create and manage atlas layouts, which are used to create multiple maps based on a single data source. Atlas layouts are useful for creating maps for large areas or for creating maps for multiple features within a single data layer.
   - Layout Manager is a powerful tool that allows you to create professional-quality map layouts within QGIS. Whether you are creating a simple map for a presentation or a complex map for a publication, the Layout Manager has the features and tools you need to get the job done.

Fig. 5.24 Layout manager user interface

Chapter 6
Data Visualization

In the realm of geospatial analysis and cartography, QGIS presents an exceptional toolset that enables users to delve into the realm of both two-dimensional (2D) and three-dimensional (3D) map creation. The purpose of this chapter lies in its unwavering dedication to guiding and illuminating, as it serves as an invaluable companion on the journey toward mastering the art of fabricating captivating and immersive 3D maps. By meticulously unraveling the intricate steps and techniques, it unfurls a comprehensive roadmap, empowering and equipping you with the knowledge and skills necessary to forge your own extraordinary 3D map. To bring this learning experience to life, it gracefully merges the celestial prowess of satellite imagery and the precise digital elevation model (DEM) of the illustrious Yamanashi Prefecture, which nestles within the awe-inspiring landscapes of Japan.

6.1 Adding Data

The first step is to add your data to QGIS. This can be done by clicking on the "Add Layer" button or by dragging and dropping your data into the QGIS window. QGIS supports a variety of file formats for vector and raster data, including shapefiles, GeoTIFFs, and CSVs.

As explained in Chap. 5, now we download the raw band of all bands from Band01 to Band12 of Yamanashi Prefecture. This time change the coordinate system to UTM 54N (EPSG:32654) so it will be consistent with DEM data. Check on all "Raw" bands then click "Download" (Fig. 6.1).

F. Ramdani, *Exploring the Earth with QGIS*, Springer Remote Sensing/Photogrammetry, https://doi.org/10.1007/978-3-031-46042-5_6

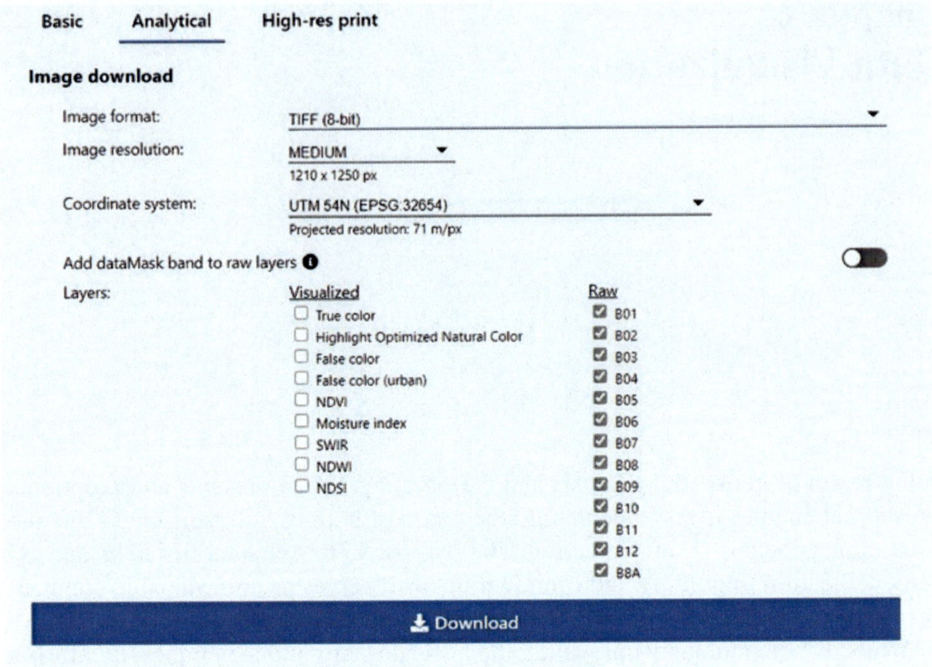

Fig. 6.1 Download analytical data

Your downloaded data will be in zip format, and then you need to unzip the file. The unzipped file will be as follow (Fig. 6.2).

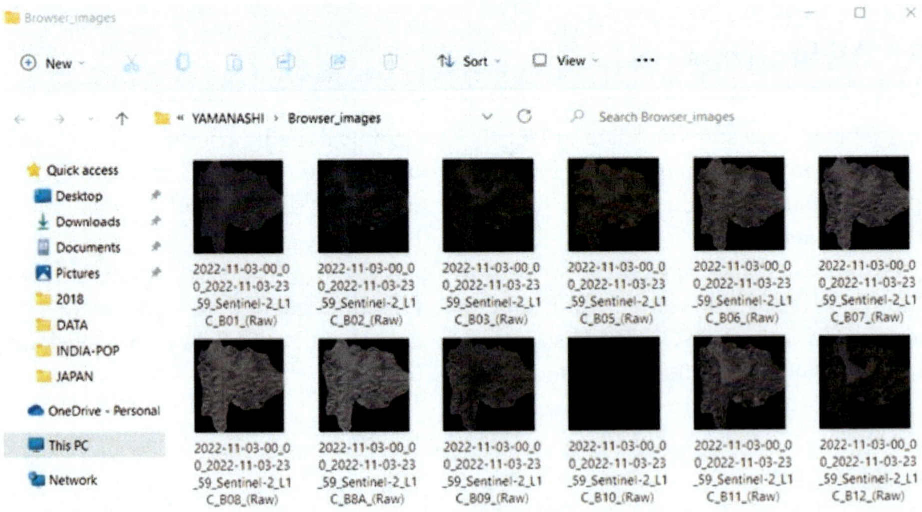

Fig. 6.2 Unzipped downloaded data

6.2 Styling

Once your data is unzipped and loaded, you can begin to style it using QGIS's styling tools. You can change the color, size, and transparency of your layers, add labels, and apply symbology to your data.

Go to Plugin > Manage and Install Plugins… > type "semi" in the search window of Plugin. Then select "Semi-Automatic Classification Plugin" and click "Install plugin."

After the installation process is finished, click on the icon  to show the SCP plugin. Then click on "Bandset" and click on the icon  to select your unzip folder. To initiate the process, ensure that you have selected "Create virtual raster" and "Create raster of..." options, click on the Run button (Fig. 6.3) and then choose the folder where you have stored the unzipped files. The completion of the process will be indicated by a series of beeps.

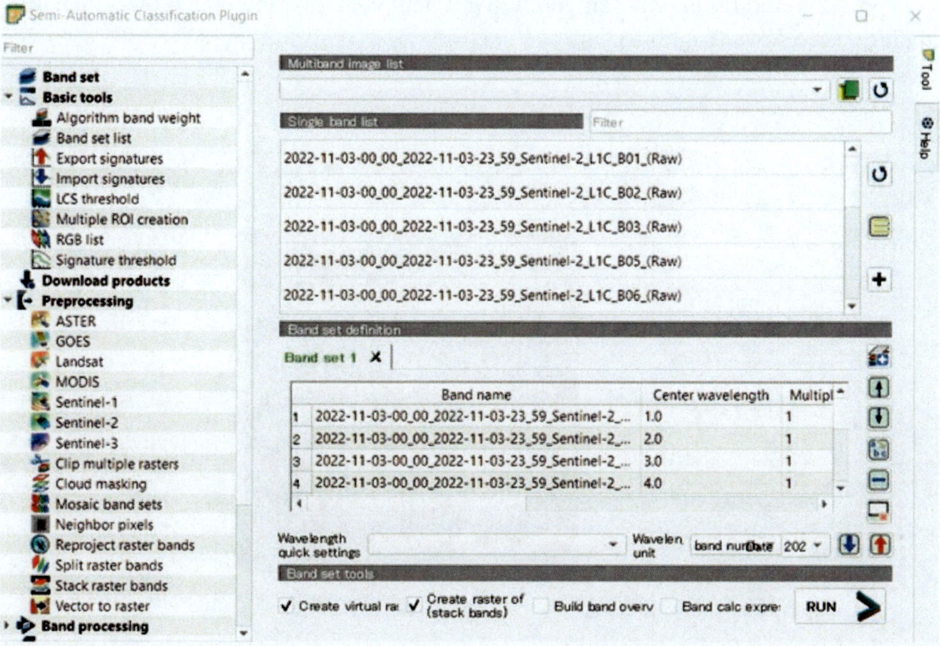

Fig. 6.3 Create band set

Look for this 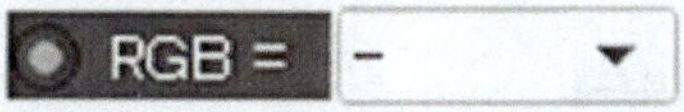 and type 4-3-2, then hit enter; the main canvas now will show you new visualization of false color as follows (Fig. 6.4).

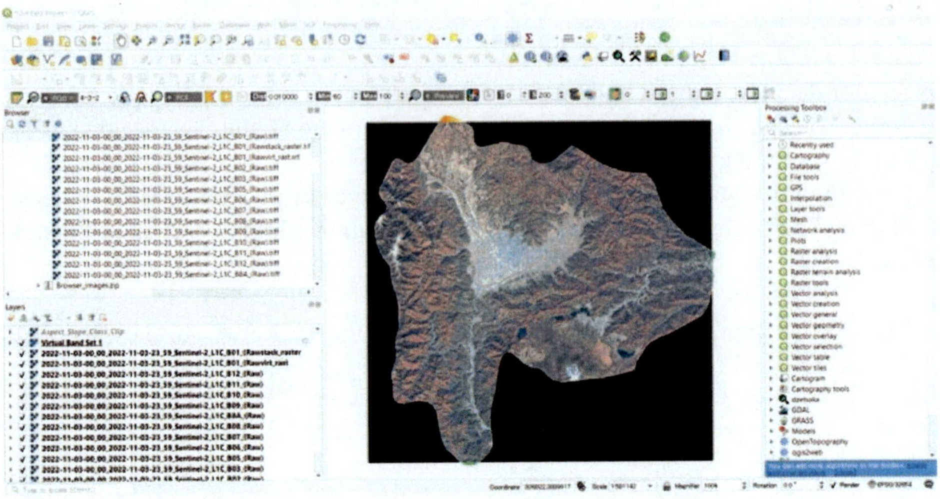

Fig. 6.4 RGB visualization

Type 3-2-1 and hit enter, then you will get new visualization that shows the natural color as follows (Fig. 6.5).

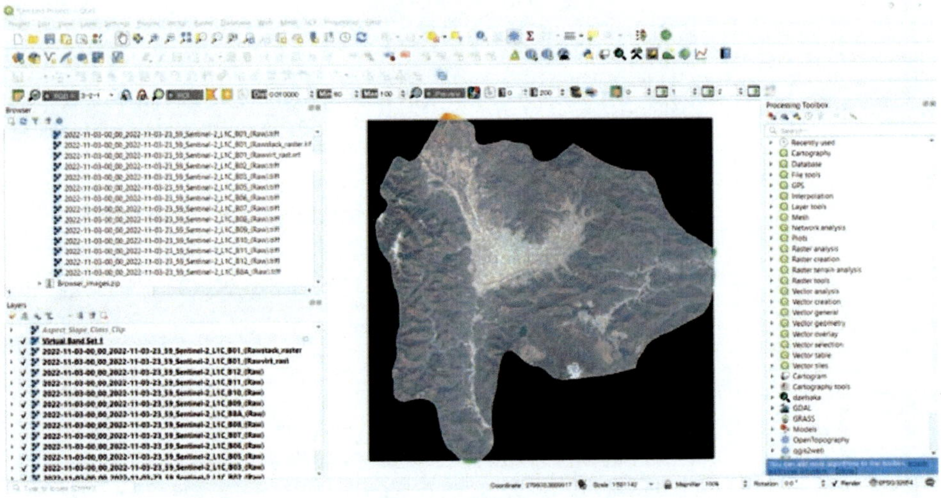

Fig. 6.5 Change the RGB visualization

Now change again the band combination to 12-9-4, we will use it for creating 3D map.

6.3 Creating 3D Maps

To create a 3D map, you can use QGIS's 3D Map View. This allows you to visualize your data in three dimensions, and you can navigate your map using your mouse or keyboard (hold Shift or Ctrl button). You can adjust the elevation and tilt your view, and you can add 3D buildings and other objects to your map.

Go to View > New 3D Map View, then click on the icon to adjust the parameters. To enhance the visualization, select "DEM (Raster Layer)" as the Type, pick your elevation file, and adjust the Vertical Scale to 2 and Tile Resolution to 32 px. However, be mindful of the Tile Resolution as a higher value would improve the visualization but might use up your computer's memory. You may also enable "Terrain Shading" before clicking OK.

To navigate, you can hold "shift" or "ctrl" on your keyboard and adjust the visualization (Fig. 6.6).

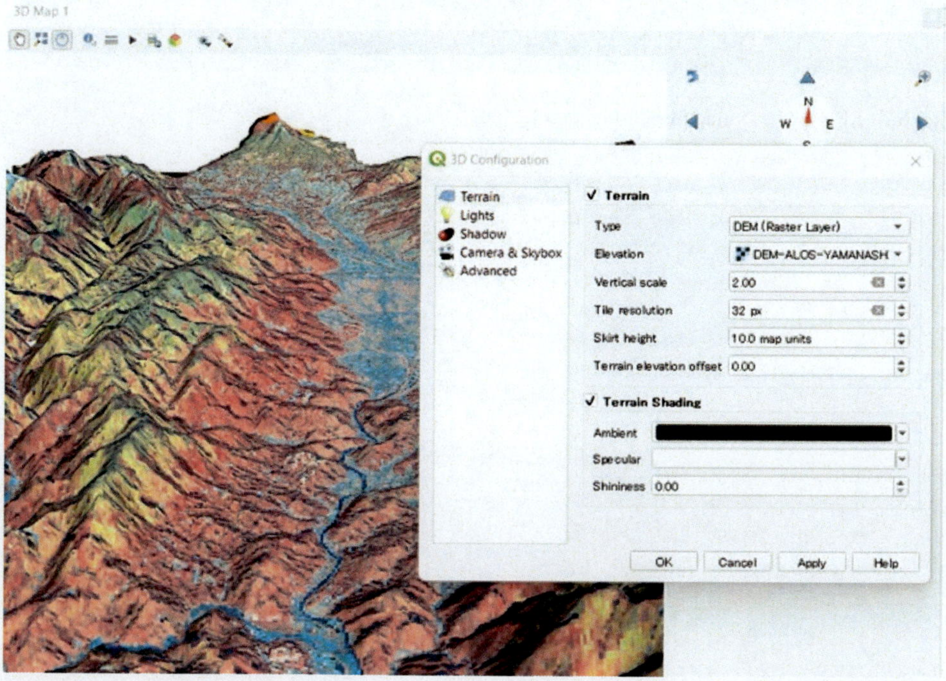

Fig. 6.6 Creating 3D map

After some adjustments, now we can see Mount Fuji from the northern part of Yamanashi Prefecture. Click on the icon if you want to save the 3D image (Fig. 6.7).

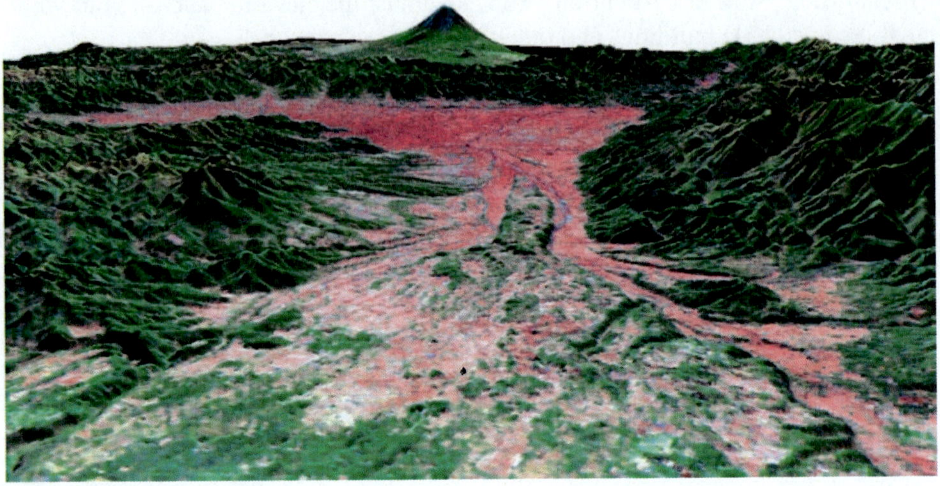

Fig. 6.7 Result of 3D map after adjustment

Chapter 7
Data Analysis

QGIS is powerful for remote sensing data analysis. It covers from basic to advance analysis. For basic image analysis, QGIS provides various basic image analysis tools, such as zooming, panning, and changing the display properties of the satellite data. You can also use the Identify Features tool to get information about the pixel values of the satellite data.

QGIS also provides various feature extraction tools, such as the SAGA GIS toolset, which includes algorithms for edge detection, segmentation, and line detection. You can also use the Orfeo Toolbox, which provides algorithms for feature extraction and classification.

Furthermore for classification using machine learning algorithms, QGIS provides several machine learning algorithms for image classification, such as Random Forest, Support Vector Machine (SVM), and k-Nearest Neighbor (k-NN). You can also use the Semi-Automatic Classification Plugin (SCP) to perform image classification using Minimum Distance, Maximum Likelihood, and Spectral Angle Mapping algorithms.

For advanced image analysis, QGIS also provides advanced image analysis tools, such as the Image Analysis Toolbar, which includes algorithms for image differencing, vegetation index calculation, and topographic analysis.

7.1 Elevation Profile Extraction

Elevation profile extraction is the process of extracting the elevation data along a specific path or route. This is often done in the context of mapping or surveying terrain, such as when creating a hiking or biking trail map or when planning a construction project in a mountainous area.

F. Ramdani, *Exploring the Earth with QGIS*, Springer Remote Sensing/Photogrammetry, https://doi.org/10.1007/978-3-031-46042-5_7

The elevation data is typically extracted from a digital elevation model (DEM), which is a dataset that represents the surface of Earth as a grid of elevation values. DEMs are typically created using satellite or aerial imagery or by using laser or radar technology to measure the distance between the ground and the sensor.

To extract an elevation profile, a specific path or route is defined, and the elevation values along that path are extracted from the DEM. This can be done using software tools that are specifically designed for elevation profile extraction, or it can be done manually by tracing the path on a map and looking up the elevation values for each point.

Once the elevation data has been extracted, it can be used to create a profile or cross-section of the terrain, which shows the elevation changes along the path. This can be useful for visualizing the steepness of a hill or mountain, identifying areas that may be prone to landslides or other hazards, or determining the best route for a hiking or biking trail.

The Profile Tool plugin is a versatile tool that enables the extraction of elevation data from DEM, either by using a temporary polyline or a selected layer. In the example shown in the figure below, the elevation of Yamanashi's DEM is extracted using a temporary polyline (Fig. 7.1).

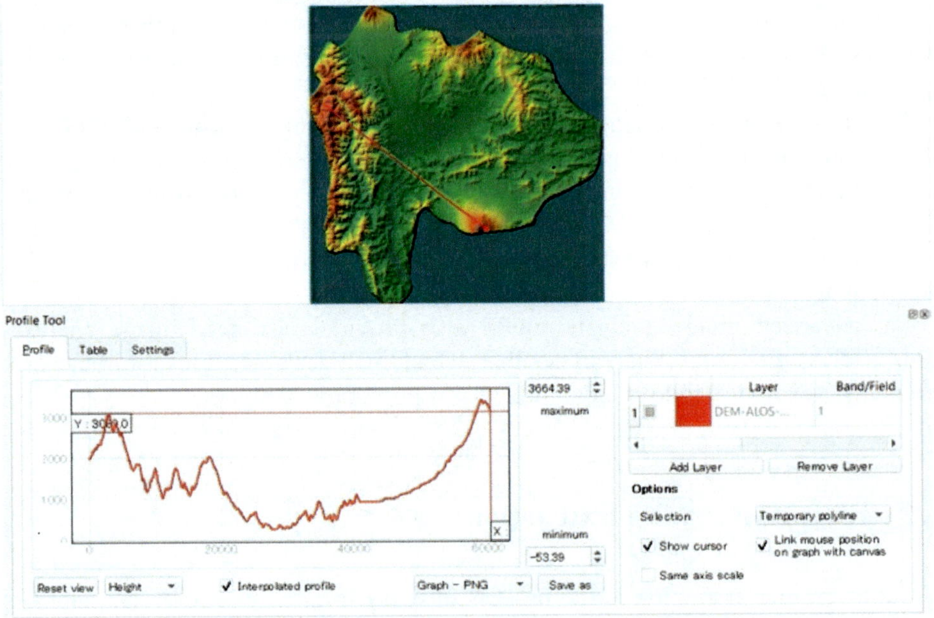

Fig. 7.1 Elevation profile extraction

To copy the result, click on "Table" and then select either "Copy to clipboard" or "Copy to clipboard (with coordinates)" (Fig. 7.2). Afterward, open your spreadsheet software and paste the result.

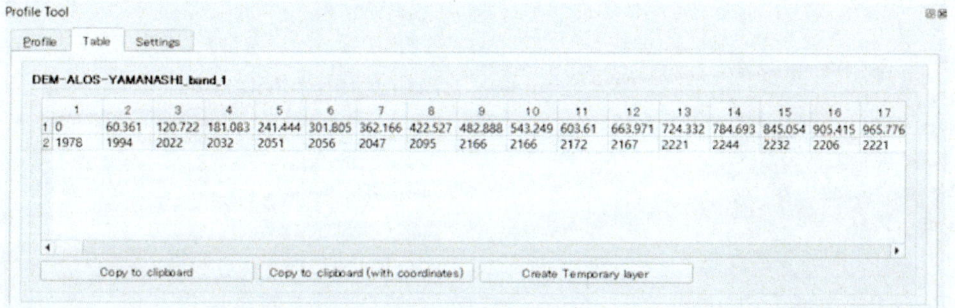

Fig. 7.2 Result of elevation profile extraction

7.2 Point Sampling Analysis

Point sampling analysis is a common method used to extract elevation values from a DEM at specific locations or points. This analysis can be useful in many applications, such as assessing slope, terrain ruggedness, and topographic position.

The first step is to determine the location of the points at which you want to extract elevation values. These points could be randomly selected or could represent specific features, such as stream gauges or vegetation plots. The next step is to load the DEM data into a QGIS.

Extract elevation values at each point by overlaying the point locations onto the DEM and querying the elevation value at that location. Once you have extracted the elevation values at each point, you can analyze them to gain insights into the terrain characteristics. For example, you could calculate the average elevation, the range of elevations, or the standard deviation of elevations at the points. These statistics can help you understand the terrain roughness or variability.

Finally, you can visualize the elevation values and any derived statistics using graphs, maps, or other visualization tools. This can help you communicate your results to others or identify patterns that may be difficult to see in tabular form.

Point sampling analysis using a DEM is a powerful technique for analyzing elevation data at specific points. By extracting elevation values and analyzing them, you can gain insights into the terrain characteristics and better understand the topography of a region.

However, with the point sampling plugin, we can not only extract the elevation values but also any other information stored in the pixel of raster data. For instance, if we have multitemporal raster data of malaria cases, we can extract this valuable information as well. First, install the Point Sampling Tool plugin (Fig. 7.3).

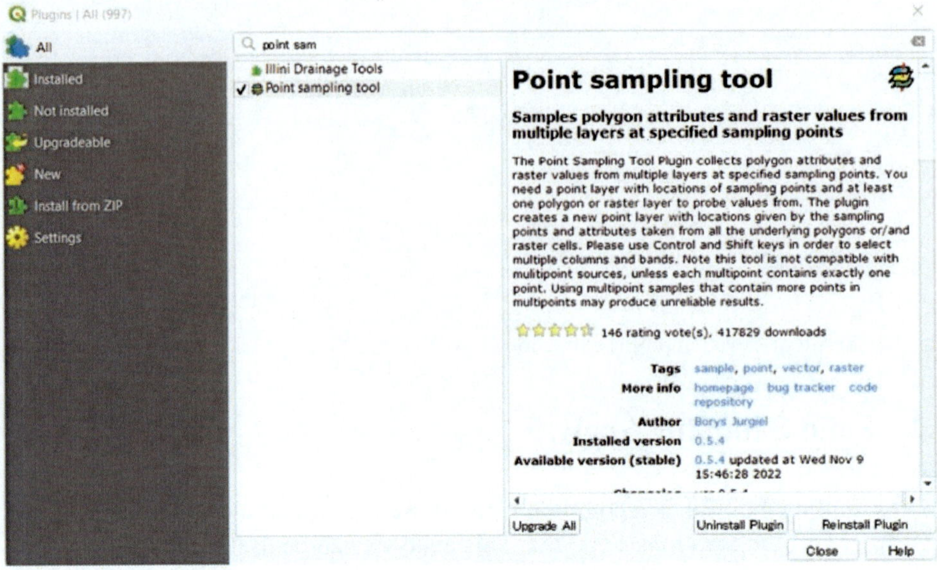

Fig. 7.3 Install point sampling tool

Then download the malaria cases from https://data.malariaatlas.org/maps?layers=Malaria:202206_Global_Pf_Incidence_Rate (Fig. 7.4).

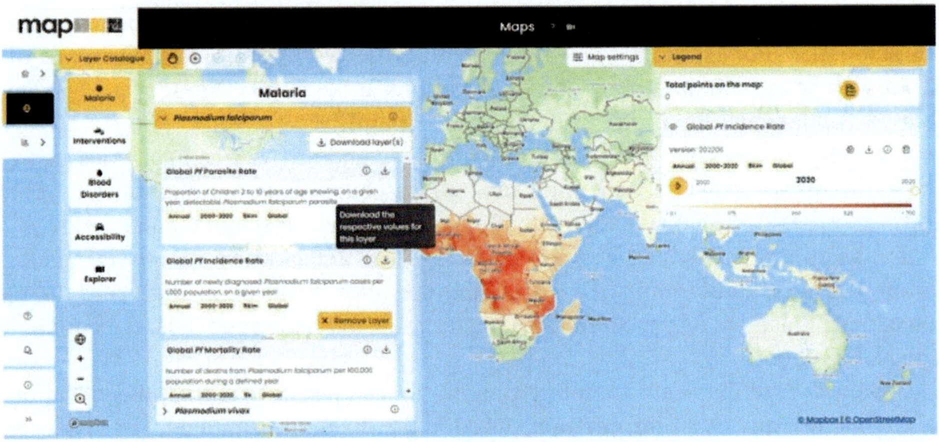

Fig. 7.4 Malaria atlas

Query only for Angola (Fig. 7.5).

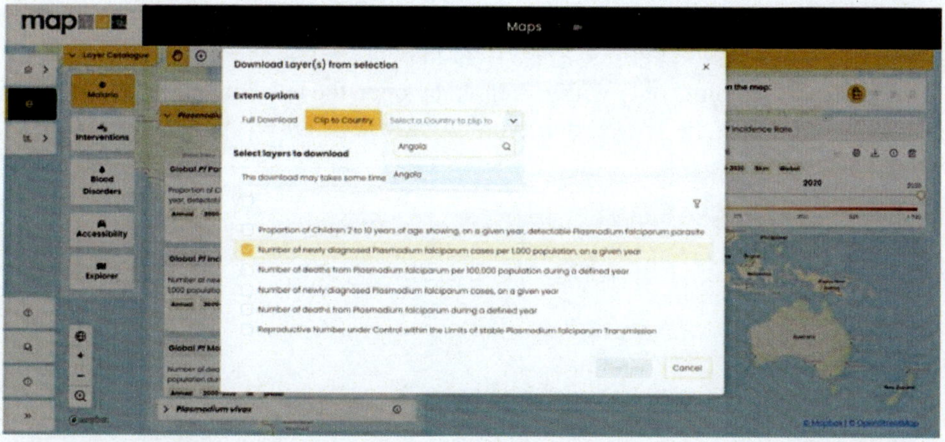

Fig. 7.5 Query data only for Angola

After completing the download, extract the data and open all using QGIS.

For points used as sampling, we will use data "ne_10m_populated_places" from Natural_Earth_quick_start. Download the data from https://www.naturalearthdata.com/downloads/ After the download is completed, click on "Natural Earth quick start kit" (Fig. 7.6) to extract the file.

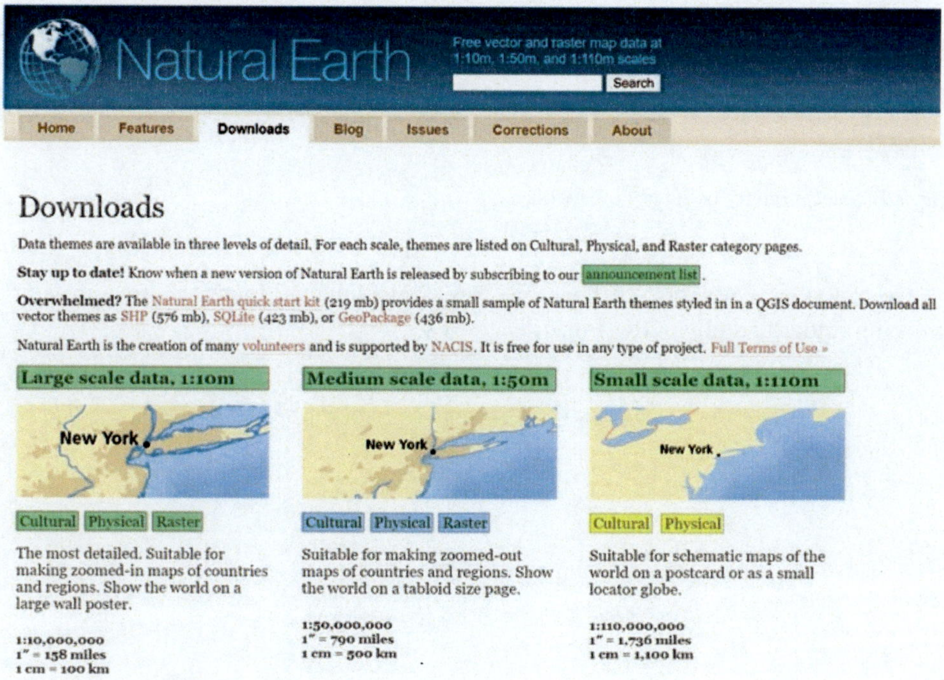

Fig. 7.6 Download Natural Earth quick start kit

On your QGIS browser, go to Natural_Earth_quick_start > packages > Natural_Earth_quick_start >10m_cultural > ne_10m_populated_places.

Then query the points only within Angola by open the attribute table , and click on icon . In the Expression type "ADM0NAME" = "Angola," then click Select Features (Fig. 7.7).

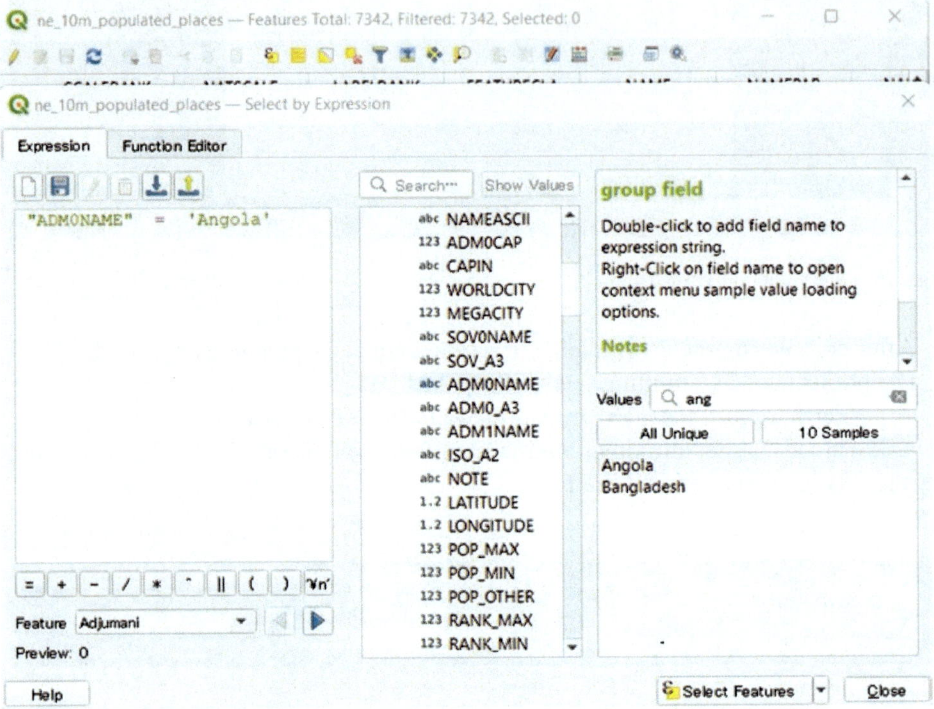

Fig. 7.7 Select country boundary of Angola

Go to Layer > Save as… Check on "Save only selected features," save in your working folder and click OK (Fig. 7.8).

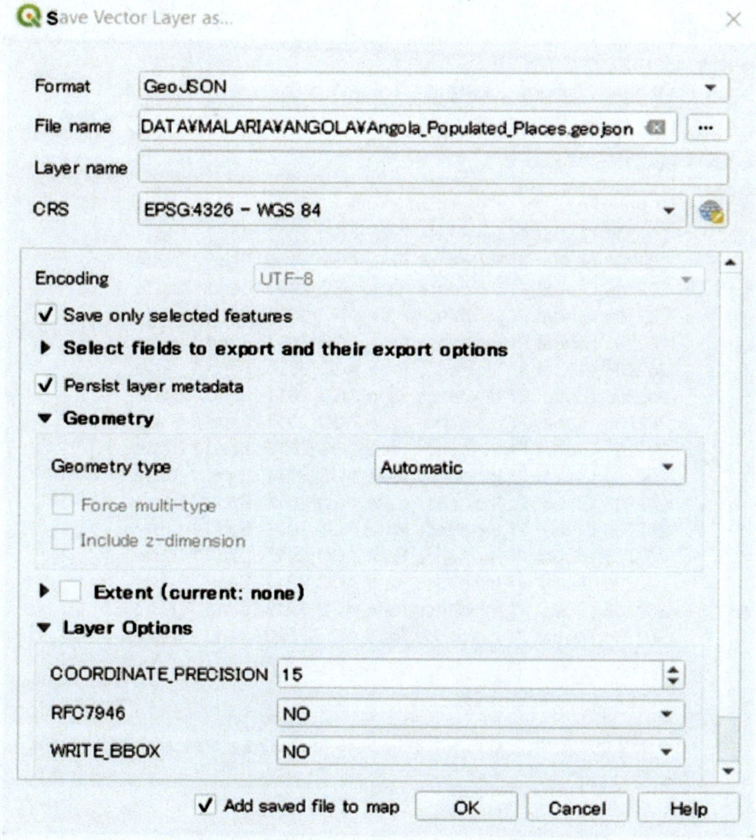

Fig. 7.8 Save the country boundary

Now click on Point Sampling Tool icon, select the "Angola_Populated_Places" as "Layer containing sampling points:" and select all malaria raster layer as "Layer with fields /bands to get values from:" by **holding Ctrl while clicking the layers' names** (Fig. 7.9). Click on Browse and save to your working folder.

Fig. 7.9 Extracting malaria cases using point sampling tool

The result will be visualized in the main canvas, then click on the resulted layer and then open the attribute table (Fig. 7.10) by click on icon  .

Angola_Populated_Malaria — Features Total: 49, Filtered: 49, Selected: 0

	fid	al_Pf_Incidence_Ra	al_Pf_Incidence_Ra	al_Pf_Incidence_Ra	al_Pf_Incidence_Ra	al_Pf_Incidence_Ra	al_Pf_Incidence_Ra	al_Pf_Incidence_Ra
1	1	NULL	NULL	NULL	NULL	NULL	NULL	NULL
2	2	0.28768	0.28102	0.2642	0.27267	0.28729	0.31144	0.31161
3	3	0.31704	0.32123	0.30742	0.31878	0.33073	0.34468	0.32237
4	4	0.18536	0.17817	0.16534	0.17034	0.17836	0.19063	0.18978
5	5	0.20757	0.20158	0.1906	0.20567	0.22915	0.25896	0.26317
6	6	0.19864	0.19449	0.18063	0.18516	0.19223	0.20336	0.19234
7	7	0.24852	0.25788	0.24494	0.24331	0.23244	0.22484	0.20152
8	8	0.36217	0.3635	0.34091	0.3306	0.3061	0.28566	0.25773
9	9	0.35499	0.36091	0.34799	0.36308	0.37406	0.39339	0.39326
10	10	0.37637	0.38298	0.37333	0.39399	0.41805	0.45199	0.45956
11	11	0.41814	0.43254	0.43036	0.4648	0.50196	0.53527	0.53879
12	12	0.3123	0.31306	0.29655	0.30548	0.31231	0.32375	0.32115
13	13	0.40571	0.4092	0.39219	0.39922	0.40215	0.40411	0.37359
14	14	0.39179	0.41078	0.40913	0.4389	0.46884	0.50083	0.4956
15	15	0.26737	0.28306	0.27037	0.27484	0.26679	0.26867	0.26493
16	16	0.36672	0.39416	0.36712	0.33629	0.26855	0.21495	0.18952

Show All Features

Fig. 7.10 Table of malaria cases of Angola

7.3 Viewshed Analysis

Viewshed analysis is a spatial analysis technique used to determine the visibility of an area from a given location or set of locations. It is a widely used technique in GIS and is based on DEM. In viewshed analysis, the DEM is used to create a raster surface of the surrounding terrain, and the location of an observer is specified as a point or a set of points. The analysis then calculates which areas are visible from the observer's location(s), based on the terrain characteristics of the area.

The calculation process of viewshed analysis involves the following steps:

- Define the observer location: The observer's location is defined by its coordinates (latitude, longitude, and altitude).
- Select the elevation surface: The DEM is used to create a raster surface of the surrounding terrain.
- Define the observer's height: The observer's height is added to the elevation surface to simulate the observer's viewpoint.
- Determine the areas visible from the observer's location: For each cell on the elevation surface, the viewshed analysis calculates whether it is visible from the observer's location. This is done by analyzing whether the line of sight between the observer and the cell is obstructed by any terrain features, such as hills or mountains.

- Output the results: The output of the viewshed analysis is typically a raster layer where each cell has a value indicating whether it is visible (1) or not visible (0) from the observer's location.

Viewshed analysis has many applications, such as in urban planning, forestry, and military operations. It is particularly useful in determining the visibility of potential development sites, identifying areas with high scenic value, identifying the maximum flying altitude for unmanned aerial vehicles (UAV), and locating observation posts or surveillance cameras.

Open QGIS and type in Processing Toolbox "viewshed" (Fig. 7.11), then there are many options you can use such as "Viewshed" under Raster miscellaneous or under Geomorphometric Analysis if you have already installed WhiteboxTools plugin (See Sect. 8.2 on how to install).

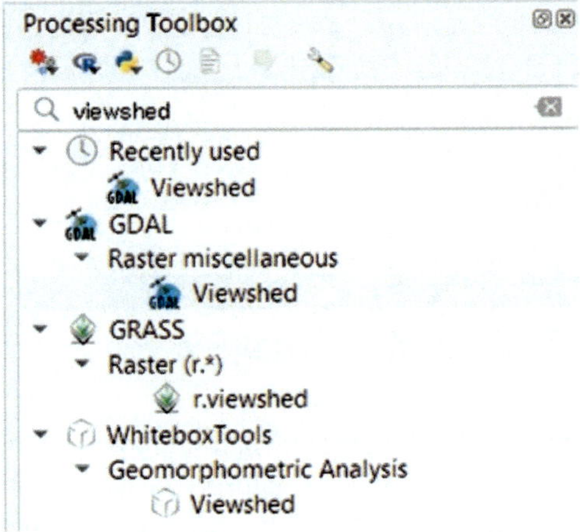

Fig. 7.11 Look for "viewshed" function in Processing Toolbox

In this case we will use "viewshed" feature under Raster miscellaneous. Set the DEM Yamanashi as "Input Layer," click on the icon  next to "Observer location" to select the point location. Set the "Observer height," "Target height," and "Maximum distance from…" (Fig. 7.12).

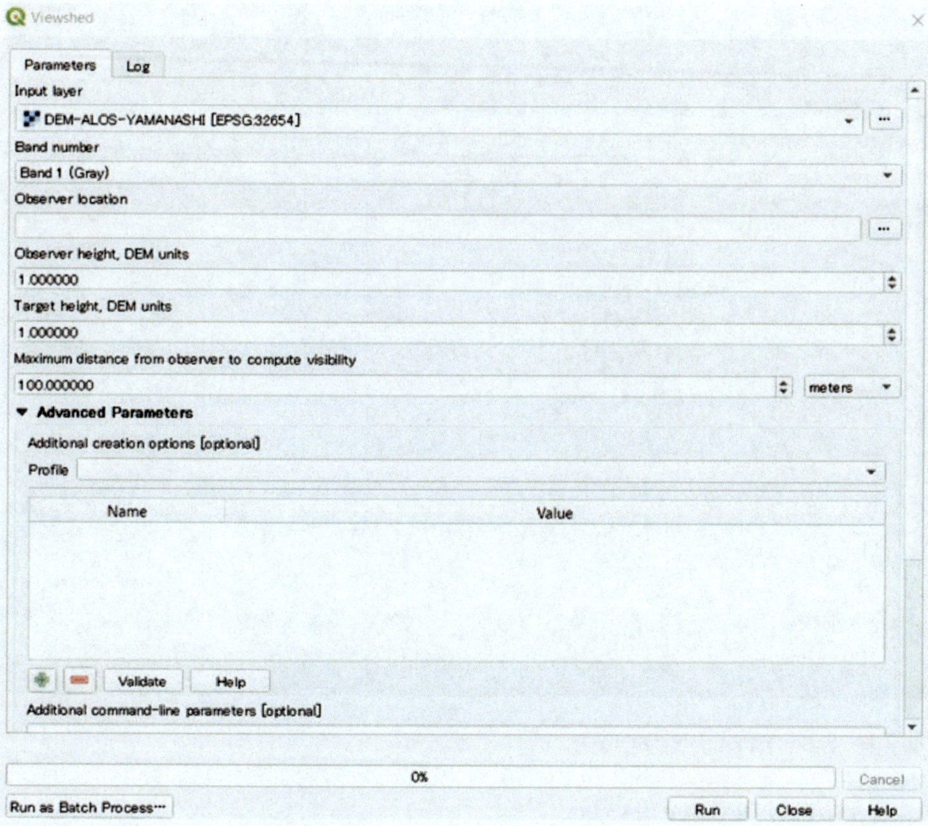

Fig. 7.12 Window of viewshed function

After clicking run, the result will be visualized in the main canvas of QGIS as shown in Fig. 7.13. The parameters used are as follows:

Observer location: center of Yamanashi.
Observer height: 1.7 m.
Target height: 100 m, and.
Maximum distance from observer to compute visibility: 10 km.

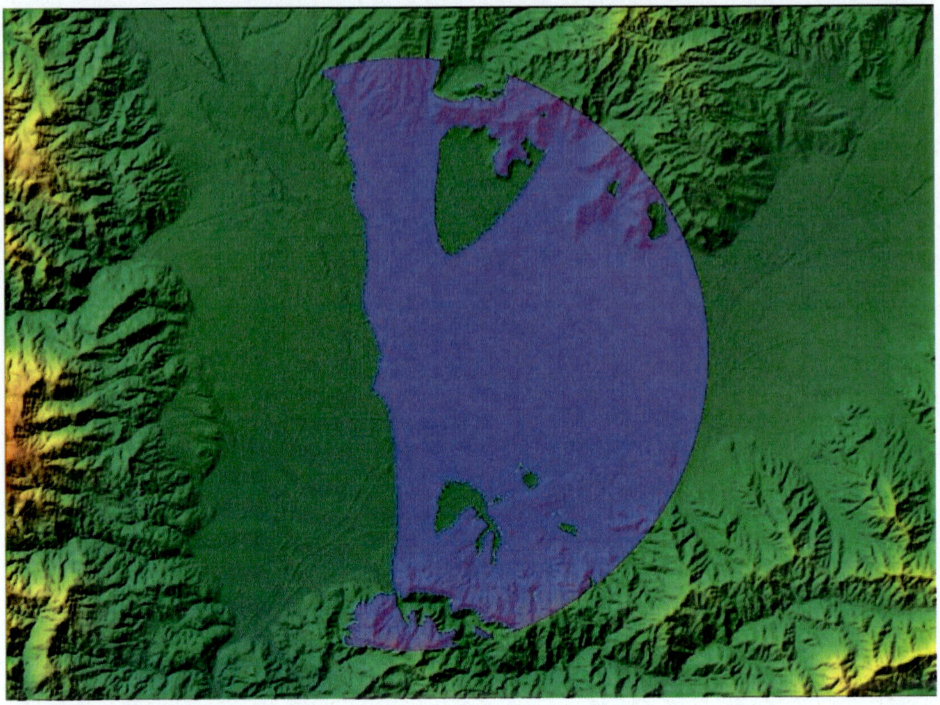

Fig. 7.13 Result of viewshed analysis

7.4 Watershed Analysis

A watershed is an area of land where all the water that falls in it and drains off of it goes into a common outlet, such as a river or a lake. Stream delineation and watershed delineation are important tasks in hydrological analysis, which help to understand the behavior of water flow in a watershed or river basin. A digital elevation model (DEM) is a widely used tool for stream and watershed delineation.

7.4.1 Stream Delineation Using DEM

Stream delineation involves identifying the path of water flow and mapping it as a line on the DEM. The process of stream delineation using a DEM is based on the concept of flow accumulation, which measures the accumulated upstream contributing area for each cell in the DEM. The basic steps for stream delineation using a DEM are as follows:

- Calculate the flow direction: The flow direction of each cell is determined by calculating the steepest descent direction to the neighboring cells.
- Calculate the flow accumulation: The flow accumulation is calculated by summing the number of cells that contribute to the flow for each cell in the DEM.

- Threshold the flow accumulation: A threshold value is used to determine the minimum flow accumulation required for a cell to be considered part of a stream. Typically, a threshold value of 100–500 is used.
- Map the stream network: The stream network is mapped by tracing the path of the cells with flow accumulation above the threshold value.

7.4.2 Watershed Delineation Using DEM

Watershed delineation involves identifying the area that contributes runoff to a particular stream segment. The process of watershed delineation using a DEM is based on the concept of flow direction, which is used to determine the direction of water flow from each cell in the DEM. The basic steps for watershed delineation using a DEM are as follows:

- Identify the outlet point: The outlet point is the point where the stream network ends and the watershed area begins.
- Calculate the flow direction: The flow direction is calculated as described above.
- Trace the flow paths: The flow paths are traced from the outlet point to the cells that contribute to the flow of water. This is done by following the steepest descent direction from each cell to its downstream neighbor.
- Determine the watershed area: The watershed area is determined by identifying all cells that contribute to the flow of water to the outlet point.

Stream and watershed delineation using DEMs involves the calculation of flow direction and accumulation, and the tracing of flow paths to identify the network of streams and the contributing area for each stream segment. This information is important for understanding the hydrological behavior of a watershed or river basin, which can help in water resource management and flood control.

Using the DEM of Yamanashi Prefecture, Japan, we will try to draw the stream and watershed. Since we need the SAGA, the Processing Saga NextGen Provider plugin needs to be installed first. Go to Plugin > Manage and Install Plugin > type "Saga," find the **Processing Saga NextGen Provider** plugin, then click on Install Plugin.

Next, download SAGA from https://sourceforge.net/projects/saga-gis/files/SAGA%20-%207/SAGA%20-%207.2.0/ Download only "saga-7.2.0_x64.zip" (Size 64.9 MB) **no need to install**. Extract the downloaded file into your working folder. Copy the path.

Open QGIS and go to Processing Toolbox, then click on the setting icon to open Options window. Go to Providers > SAGANG > SAGA Folder, then paste the path where you extract the SAGA (Fig. 7.14).

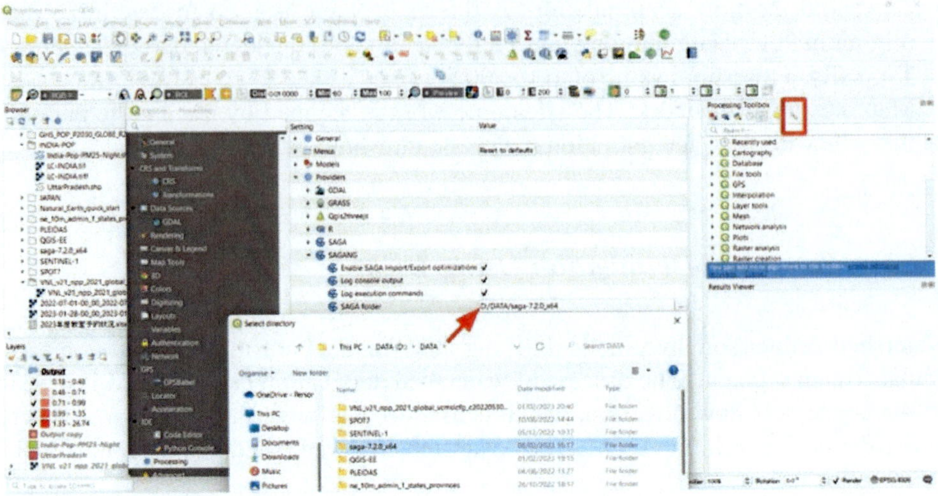

Fig. 7.14 Install Saga NextGen

Next, go to Processing Toolbox, and type "Fill sink" on the search window, double-click on "Fill sinks (Wang & Liu)" under SAGA Next Gen > Terrain Analysis – Hydrology. In DEM, select the Reprojected (wrap) data and save the result into your working folder (Fig. 7.15).

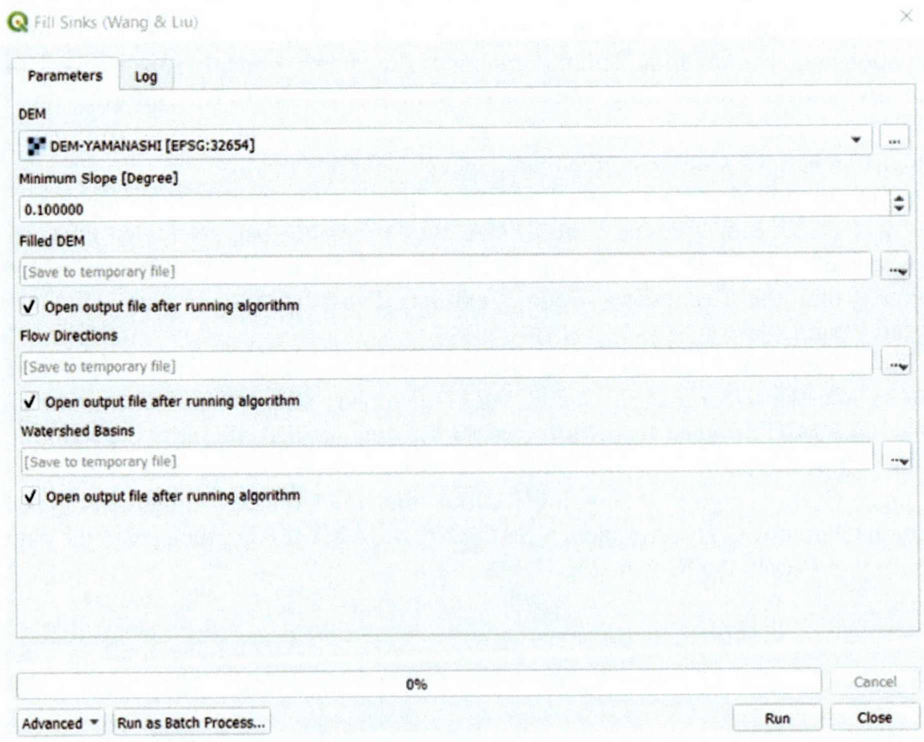

Fig. 7.15 Run the Fill and sink function

The result will be three different layers: filled DEM, watershed basin, and flow direction. Next, we will produce the stream. Type “strahler” then double-click on “Strahler order” under SAGA Next Gen > Terrain Analysis – Channels. The new window of Strahler order will open (Fig. 7.16).

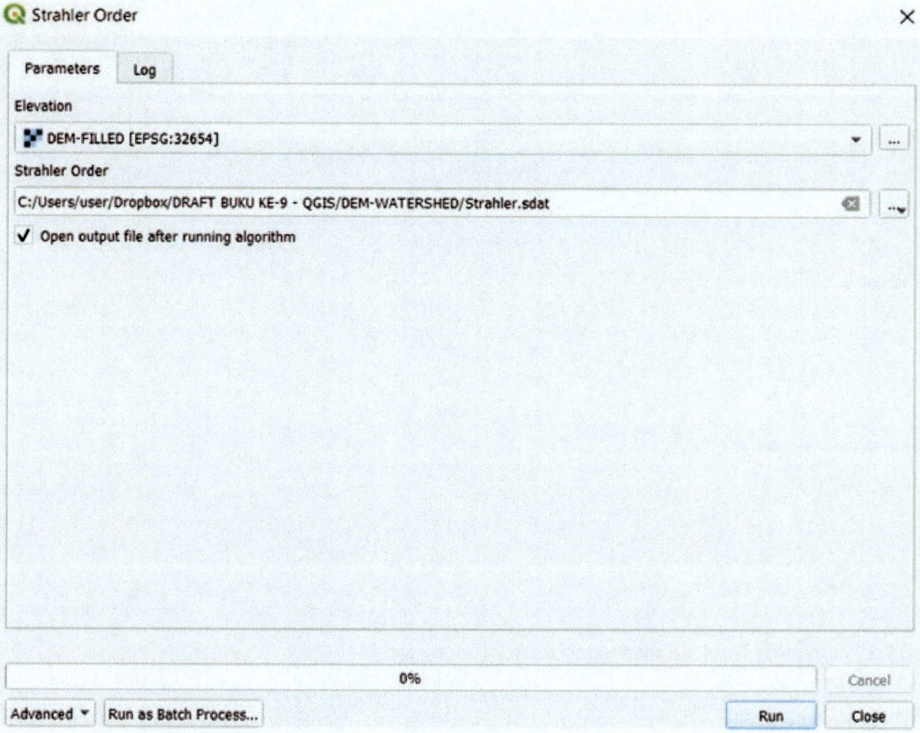

Fig. 7.16 Window of Strahler order

Next, we need to create a Boolean map with 1 (True) for Strahler order > = 5 and 0 (False) for the other values. Go to Raster > Raster Calculator (Fig. 7.17), then input the following expression:

```
"Strahler@1" >= 5
```

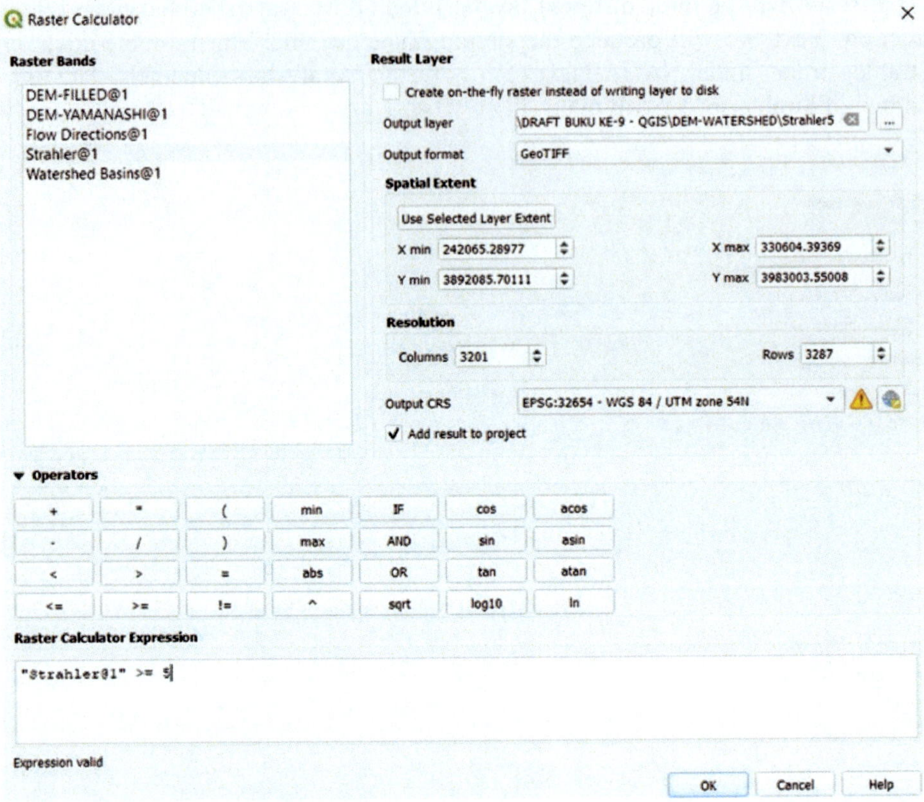

Fig. 7.17 Creating Boolean map using Raster Calculator

The next step is to convert the watershed from raster to vector, go to Raster > Conversion > Polygonize (Raster to Vector) then window of Polygonize (Raster to Vector) will open (Fig. 7.18).

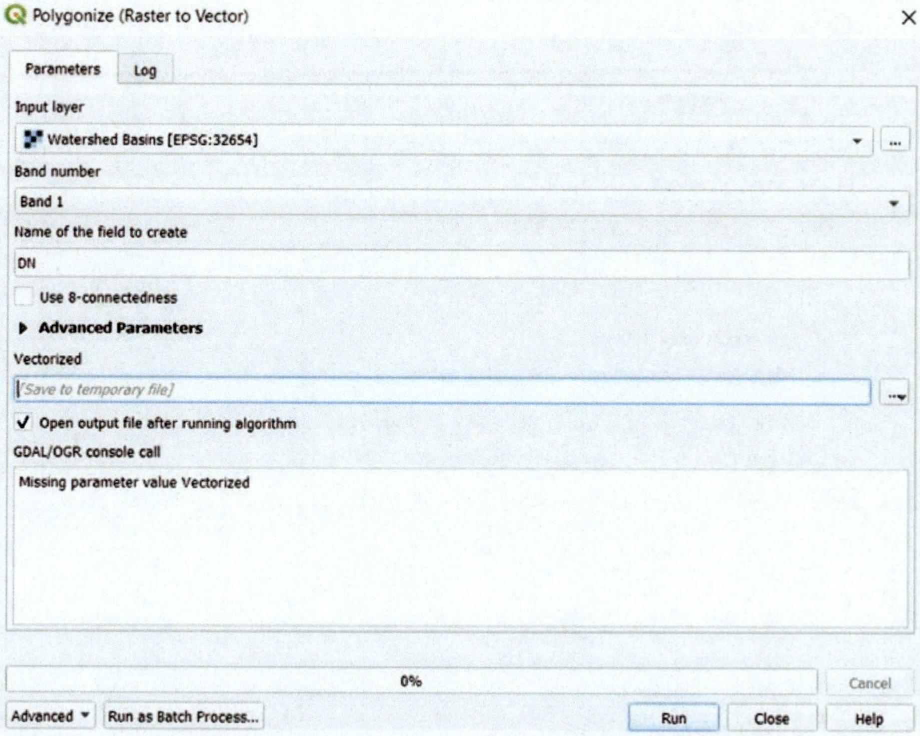

Fig. 7.18 Window of Polygonize (Raster to Vector)

Now select only the watershed in the center of Yamanashi using the Selected feature's icon , right-click and select "Save features as". Do not forget to check on "Save only selected features" (Fig. 7.19).

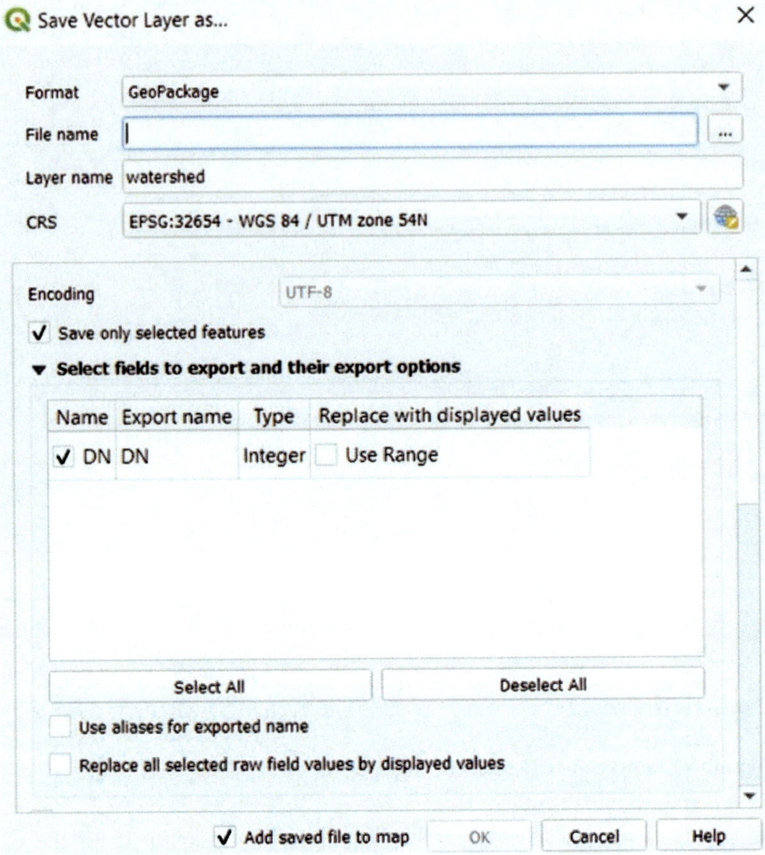

Fig. 7.19 Save the polygon as vector file

Now we can clip the "Strahler" layer using the watershed boundary, go to Raster > Extraction > Clip Raster by Mask Layer (Fig. 7.20).

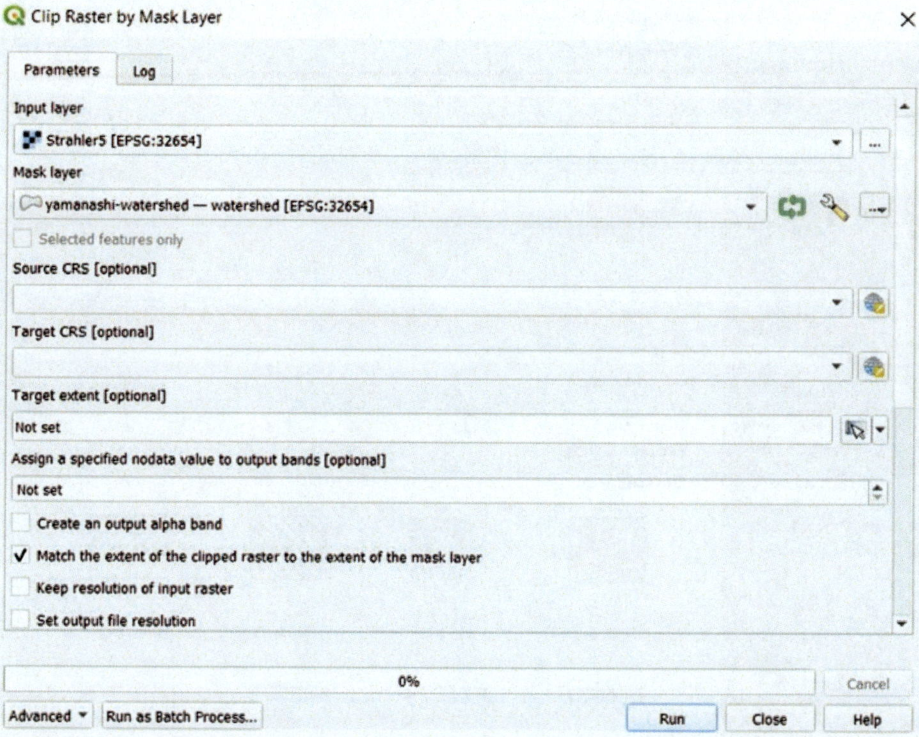

Fig. 7.20 Clip raster by mask layer

Next double-click on layer "Yamanashi-watershed," in the Symbology click on the drop-down menu and change from "Single symbol" to "Inverted polygon" (Fig. 7.21).

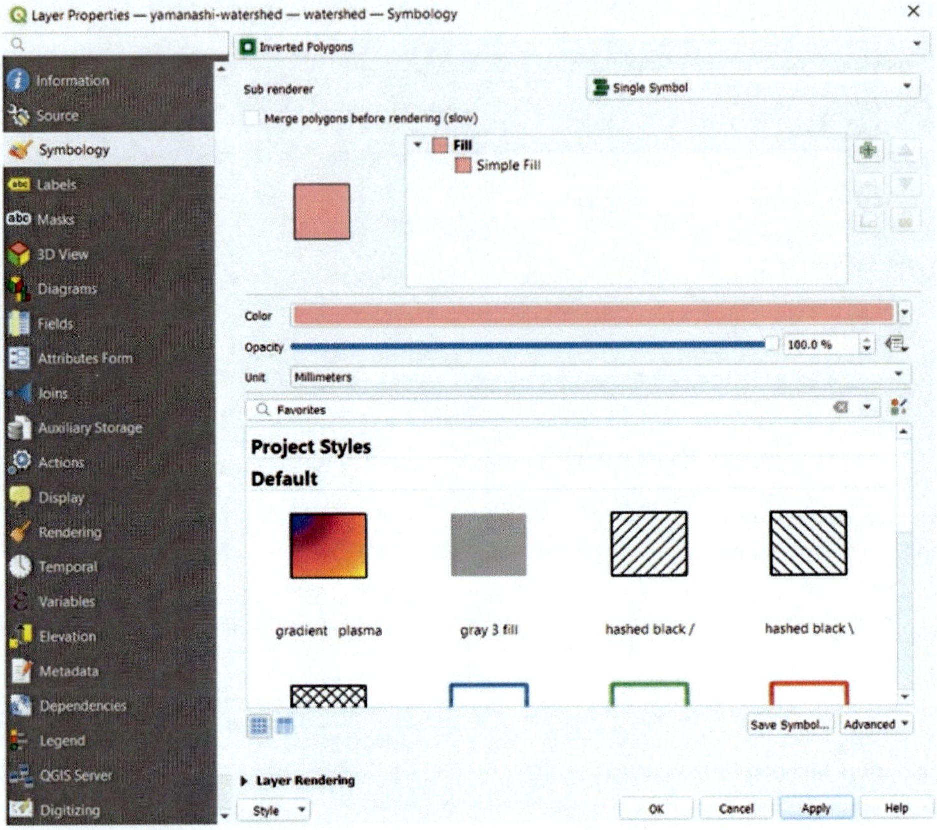

Fig. 7.21 Change the visualization of watershed

Now click on "Simple Fill" under "Fill," and change Symbol layer type to "Shapeburst" and change the parameters as shown in the Fig. 7.22. Set distance to 4.00 and Opacity level to 65%, and change color to Black.

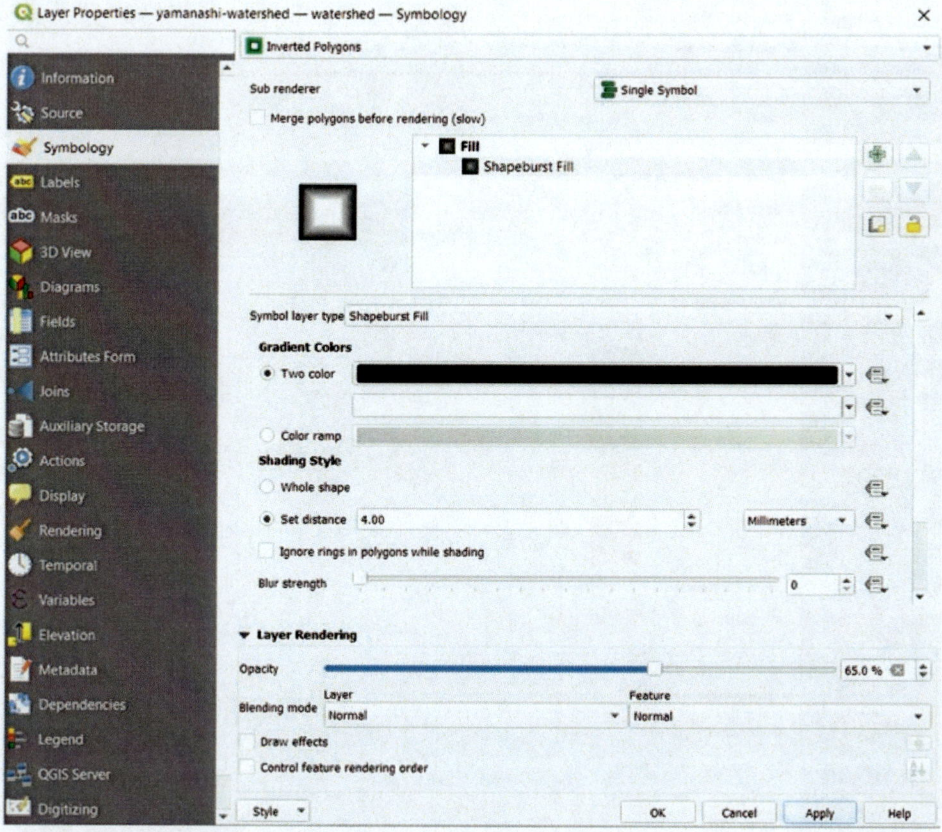

Fig. 7.22 Change the visualization parameters

Now right-click on the layer "DEM-YAMANASHI" and select "Duplicate." Double-click on the duplicate layer "DEM-YAMANASHI-copy," and change the symbology parameters as follows. Change the Render type to "Singleband pseudo-color." Click on the small arrow next to "Color ramp" then select Create new color ramp > Catalog: cpt-city. Then scroll down, click on Topography > elevation. Change the Blending mode to "Multiply" under Layer Rendering menu, then click OK (Fig. 7.23).

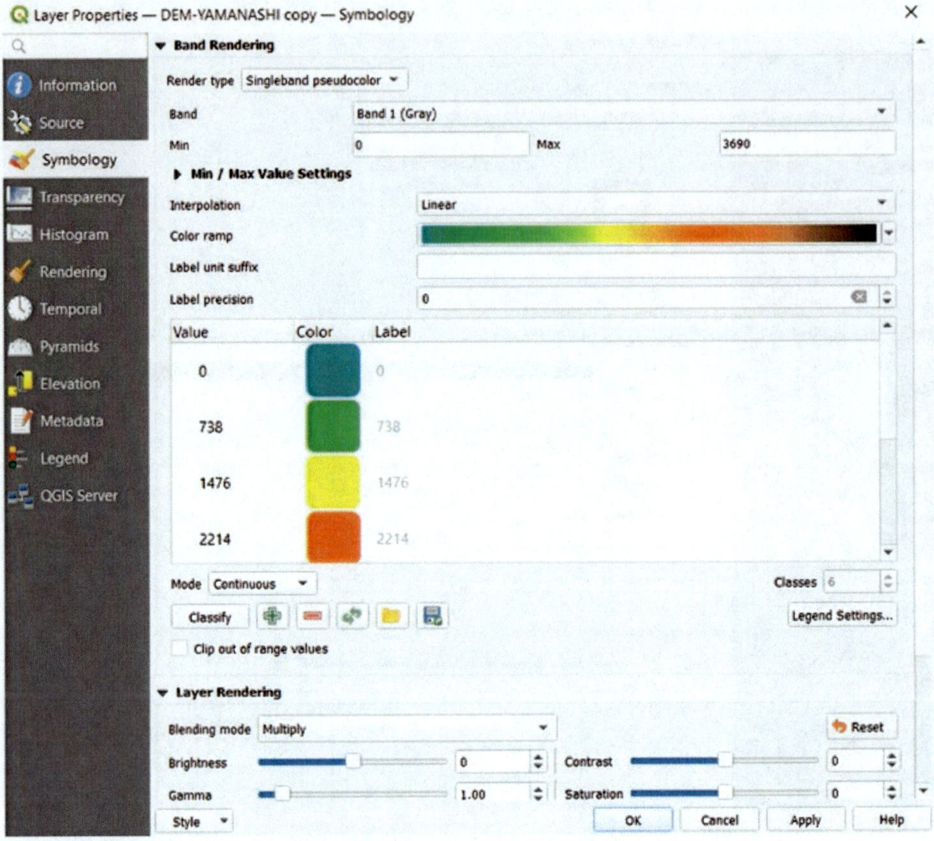

Fig. 7.23 Adjust the visualization parameters of duplicate layer

Double-click on the original "DEM-YAMANASHI" layer and change the symbology as follows. Change the Render type to "Hillshade," and then change the Blending mode to "Multiply," and click OK (Fig. 7.24).

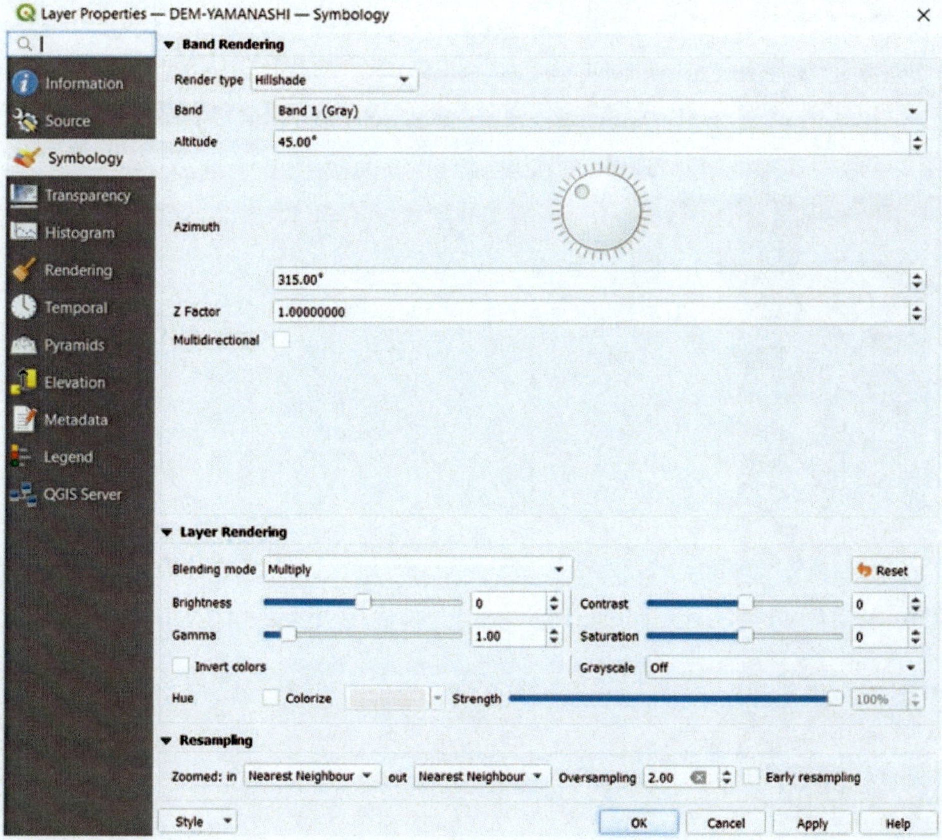

Fig. 7.24 Adjust the DEM visualization parameters

Clip the raster layer of "Strahler5" using "Yamanashi-watershed" and convert the clipped data into vector, go to Raster > Conversion > Polygonize (Raster to Vector), give it name "Streams." Then double-click on the polygonized layer, change the symbology to "Categorized" and delete all class except "1." Change the color into the blue (Fig. 7.25).

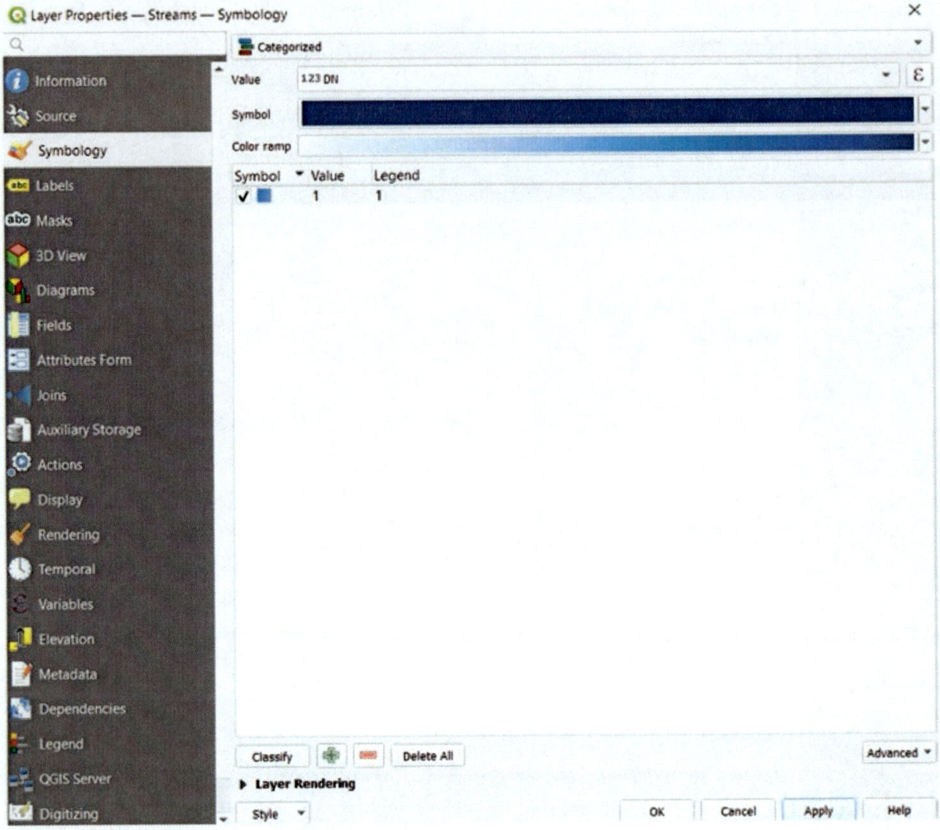

Fig. 7.25 Change the visualization of streams

Arrange the layer as shown in Fig. 7.26.

Fig. 7.26 Layer arrangement

And after some symbology adjustment, the final result will be shown as Fig. 7.27.

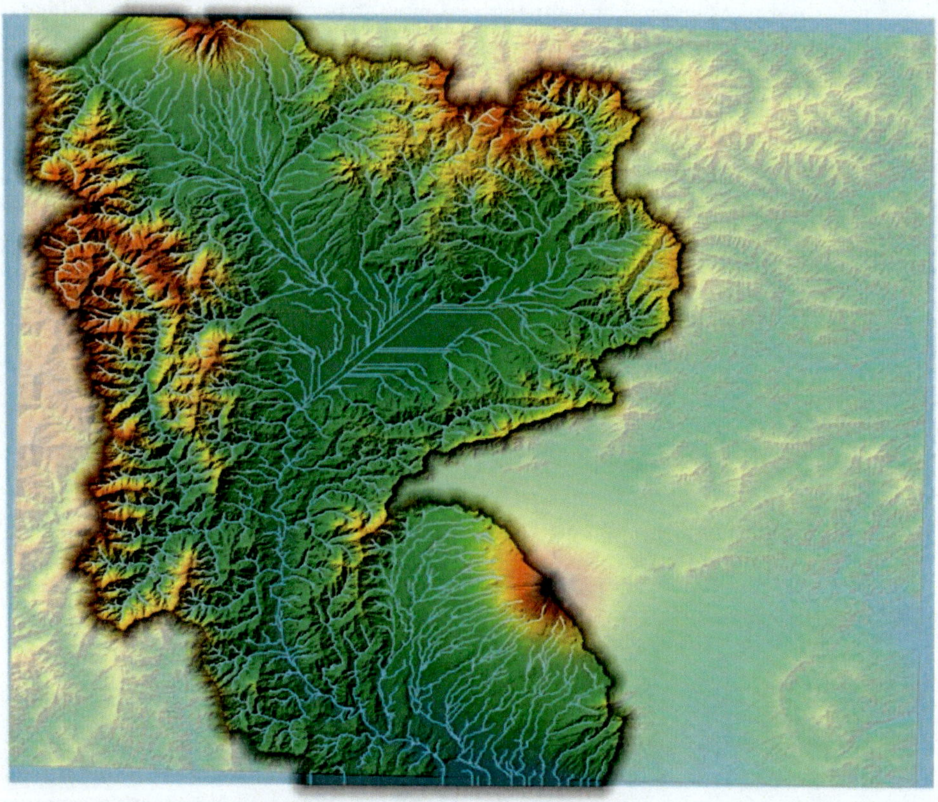

Fig. 7.27 Final visualization result of DEM overlaid with streams layer

7.5 Statistical Analysis

7.5.1 Zonal Statistics

Zonal statistics is a powerful spatial analysis technique that allows you to calculate statistics on a raster layer (such as satellite remote sensing data) for defined zones or regions (such as administrative boundaries or ecological units) in a vector layer (such as a polygon shapefile). In QGIS, you can perform zonal statistics using the "Zonal Statistics" tool from the "Raster" menu.

To perform zonal statistics using QGIS and satellite remote sensing data, you will need:

1. A satellite remote sensing raster layer that contains the data you want to analyze. This could be, for example, a layer containing information on land cover, vegetation indices, or temperature.

2. A vector polygon layer that defines the zones or regions you want to analyze. This could be, for example, a layer containing administrative boundaries, ecological units, or land cover classes.

Once you have these layers loaded into QGIS, follow these steps to perform zonal statistics:

- Open the "Zonal statistics" tool under the "Raster Analysis" menu in the Processing Toolbox (Fig. 7.28).

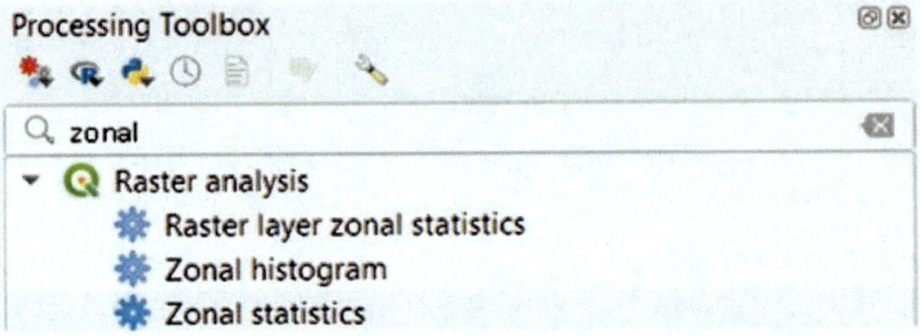

Fig. 7.28 Zonal statistic

- In the "Input layer" field, select the vector polygon layer that defines the zones or regions you want to analyze. In this case, we use India's provincial boundary.
- In the "Zonal Statistics" dialog box (Fig. 7.29), select the satellite remote sensing raster layer you want to analyze in the "Raster layer" field. In this case we will use night-time data. We use Annual VNL V21 of year 2021 for global scale (available at https://eogdata.mines.edu/nighttime_light/annual/v21/). Download only average data (VNL_v21_npp_2021_global_vcmslcfg_c202205302300.average.dat.tif.gz). After downloading, unzip and open the raster data in QGIS.

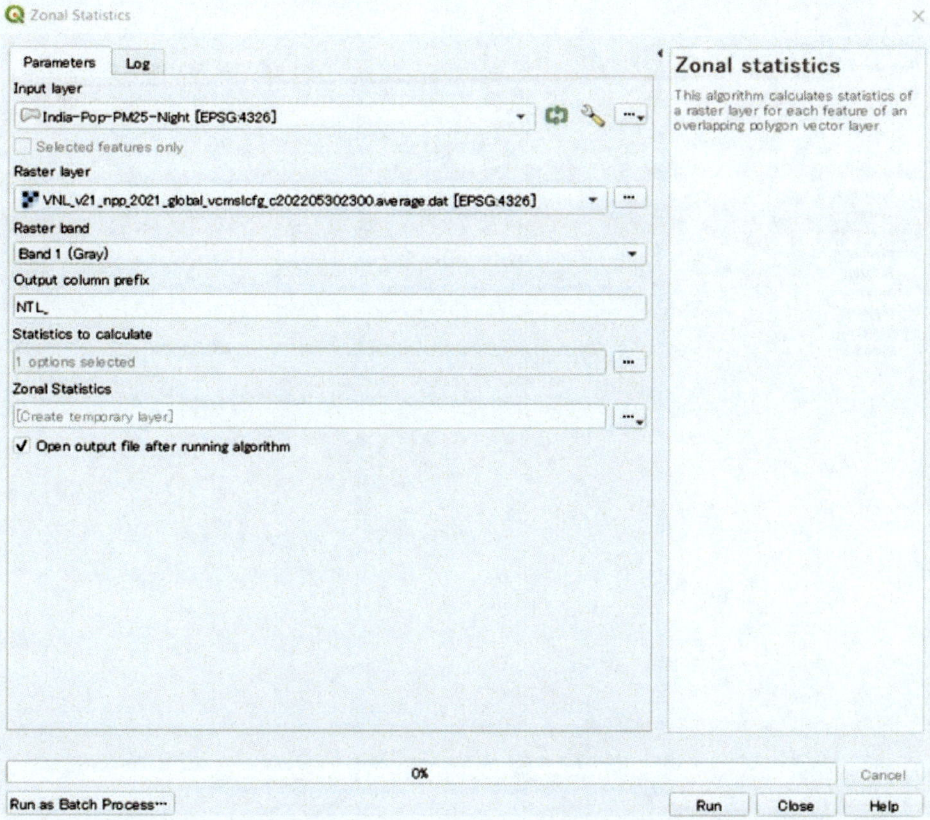

Fig. 7.29 Zonal statistic dialog box

- In the "Statistics to calculate" field, select the statistics you want to calculate (Fig. 7.30). You can choose from a range of statistics, such as mean, median, maximum, minimum, and standard deviation.

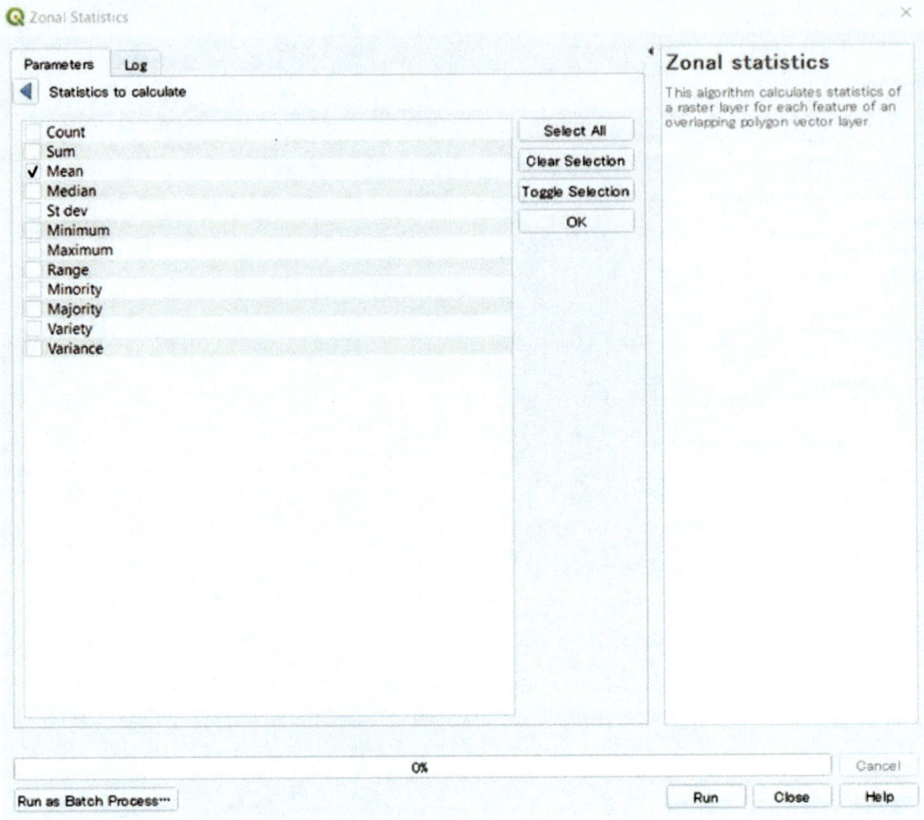

Fig. 7.30 Statistic to calculate

- In the "Output column prefix" field, enter a name for the column that will contain the calculated statistics. For example "NTL_" since we calculate the average of night-time light.
- Click "OK" to run the zonal statistics analysis.

Once the analysis is complete, you will see a new column added to the attribute table of the vector polygon layer, containing the calculated statistics for each zone or region (Fig. 7.31).

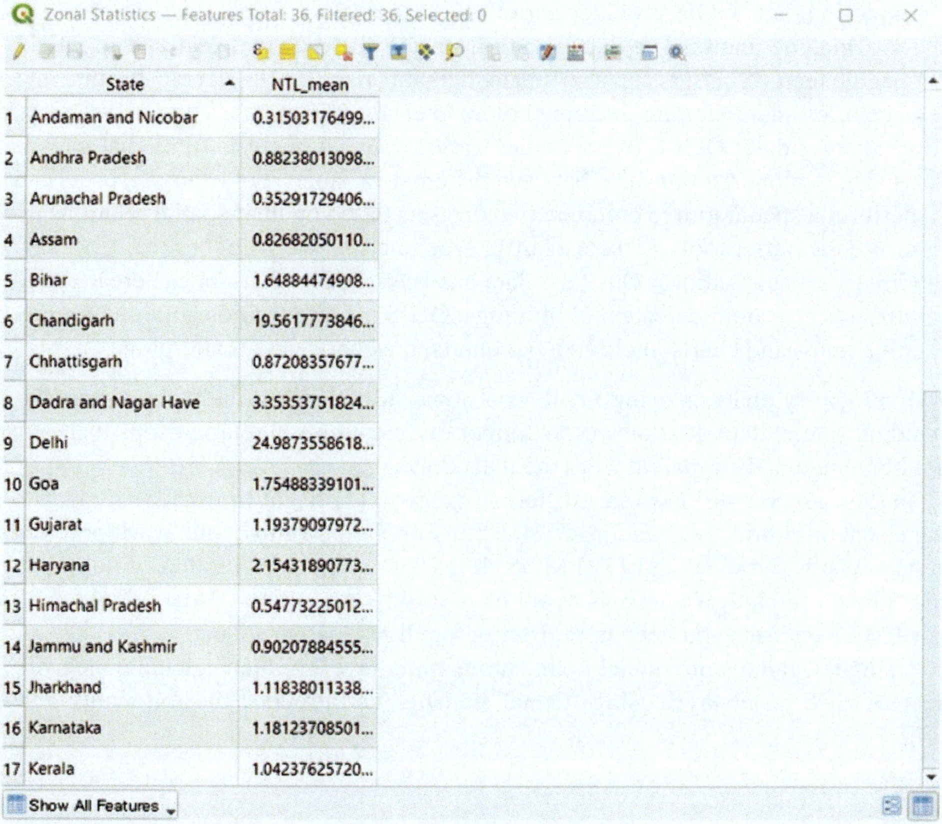

	State	NTL_mean
1	Andaman and Nicobar	0.31503176499...
2	Andhra Pradesh	0.88238013098...
3	Arunachal Pradesh	0.35291729406...
4	Assam	0.82682050110...
5	Bihar	1.64884474908...
6	Chandigarh	19.5617773846...
7	Chhattisgarh	0.87208357677...
8	Dadra and Nagar Have	3.35353751824...
9	Delhi	24.9873558618...
10	Goa	1.75488339101...
11	Gujarat	1.19379097972...
12	Haryana	2.15431890773...
13	Himachal Pradesh	0.54773225012...
14	Jammu and Kashmir	0.90207884555...
15	Jharkhand	1.11838011338...
16	Karnataka	1.18123708501...
17	Kerala	1.04237625720...

Fig. 7.31 Table of result from zonal statistics analysis

Zonal statistics using QGIS and satellite remote sensing data can be used for a range of applications, such as analyzing vegetation cover, identifying land use changes, and monitoring climate trends.

7.5.2 Exploratory Analysis

Exploratory analysis using QGIS refers to the process of visually analyzing spatial data to identify patterns, trends, and relationships. Here are some steps to perform exploratory analysis using QGIS:

- Data preparation: Before starting the analysis, it is important to gather relevant data and prepare it for analysis. This may involve cleaning, formatting, and importing data into QGIS.
- Load data into QGIS: Once the data is prepared, it can be loaded into QGIS. This can be done by selecting "Add Vector Layer" from the Layer menu and choosing the file or database where the data is stored.

- Exploring data: QGIS provides various tools to explore spatial data. Some of the tools include zooming, panning, and querying the data to select and highlight specific features. Users can also create thematic maps by classifying the data into categories and assigning different colors to each category.
- Analyzing data: QGIS also provides various spatial analysis tools that can be used to explore relationships between different variables. For example, users can perform a spatial join to combine two datasets based on their spatial relationship or use the buffer tool to create a buffer around a specific point or area.
- Creating visualizations: Once the data has been analyzed, users can create visualizations to communicate their findings. QGIS provides various options for creating maps and charts, including bar charts, pie charts, and scatterplots.

Exploratory analysis using QGIS is a powerful tool for visualizing and understanding spatial data. It allows users to identify patterns and relationships that may not be immediately apparent from the data alone.

In this case we will use three different datasets: (1) Night-time (https://eogdata.mines.edu/nighttime_light/annual/v21/), (2) Population (https://hub.worldpop.org/geodata/listing?id=32), and (3) PM2.5 (https://sedac.ciesin.columbia.edu/data/set/sdei-global-annual-gwr-pm2-5-modis-misr-seawifs-aod-v4-gl-03/data-download). For PM2.5 datasets you need to register before download the data.

Using the India's provincial boundary as input layer we then calculate the averages of each raster layer using "Zonal statistic" and produce the following table (Fig. 7.32).

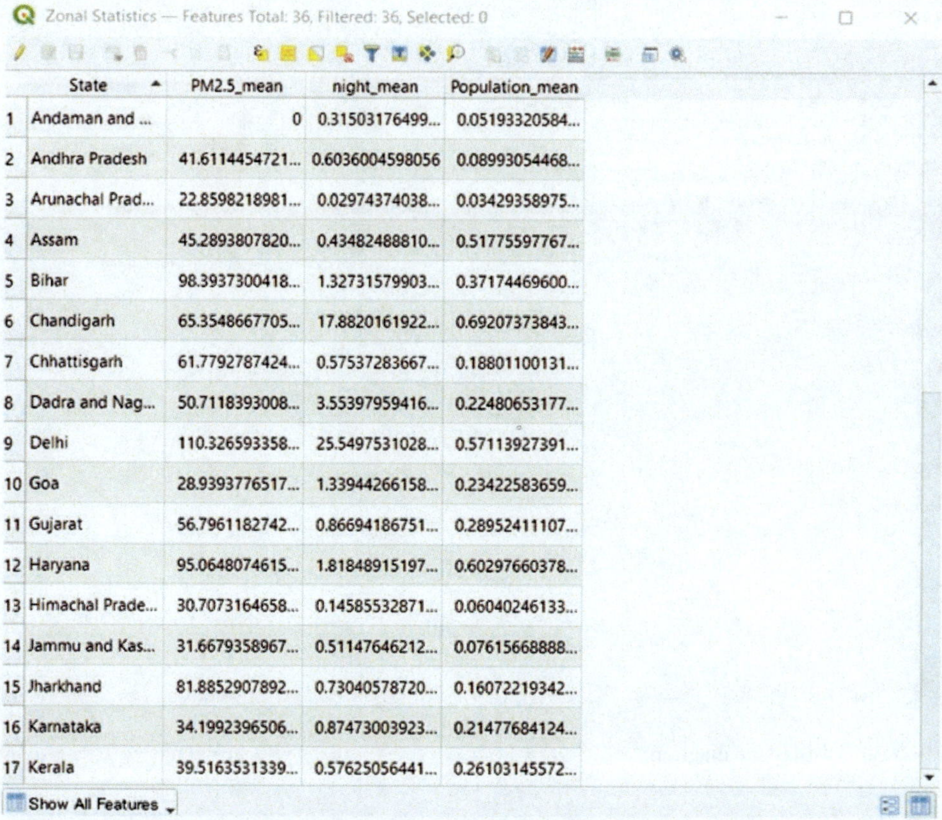

Zonal Statistics — Features Total: 36, Filtered: 36, Selected: 0

	State	PM2.5_mean	night_mean	Population_mean
1	Andaman and ...	0	0.31503176499...	0.05193320584...
2	Andhra Pradesh	41.6114454721...	0.6036004598056	0.08993054468...
3	Arunachal Prad...	22.8598218981...	0.02974374038...	0.03429358975...
4	Assam	45.2893807820...	0.43482488810...	0.51775597767...
5	Bihar	98.3937300418...	1.32731579903...	0.37174469600...
6	Chandigarh	65.3548667705...	17.8820161922...	0.69207373843...
7	Chhattisgarh	61.7792787424...	0.57537283667...	0.18801100131...
8	Dadra and Nag...	50.7118393008...	3.55397959416...	0.22480653177...
9	Delhi	110.326593358...	25.5497531028...	0.57113927391...
10	Goa	28.9393776517...	1.33944266158...	0.23422583659...
11	Gujarat	56.7961182742...	0.86694186751...	0.28952411107...
12	Haryana	95.0648074615...	1.81848915197...	0.60297660378...
13	Himachal Prade...	30.7073164658...	0.14585532871...	0.06040246133...
14	Jammu and Kas...	31.6679358967...	0.51147646212...	0.07615668888...
15	Jharkhand	81.8852907892...	0.73040578720...	0.16072219342...
16	Karnataka	34.1992396506...	0.87473003923...	0.21477684124...
17	Kerala	39.5163531339...	0.57625056441...	0.26103145572...

Show All Features

Fig. 7.32 Attribute table of zonal statistic result of three different raster layers

Then the symbology and histogram can be adjusted. Double-click on the layer and go to Diagrams. Select "Histogram" and add the attribute that want to be display in histogram. In this case we will display Population, PM2.5, and night-time average values. Change the color if needed (Fig. 7.33).

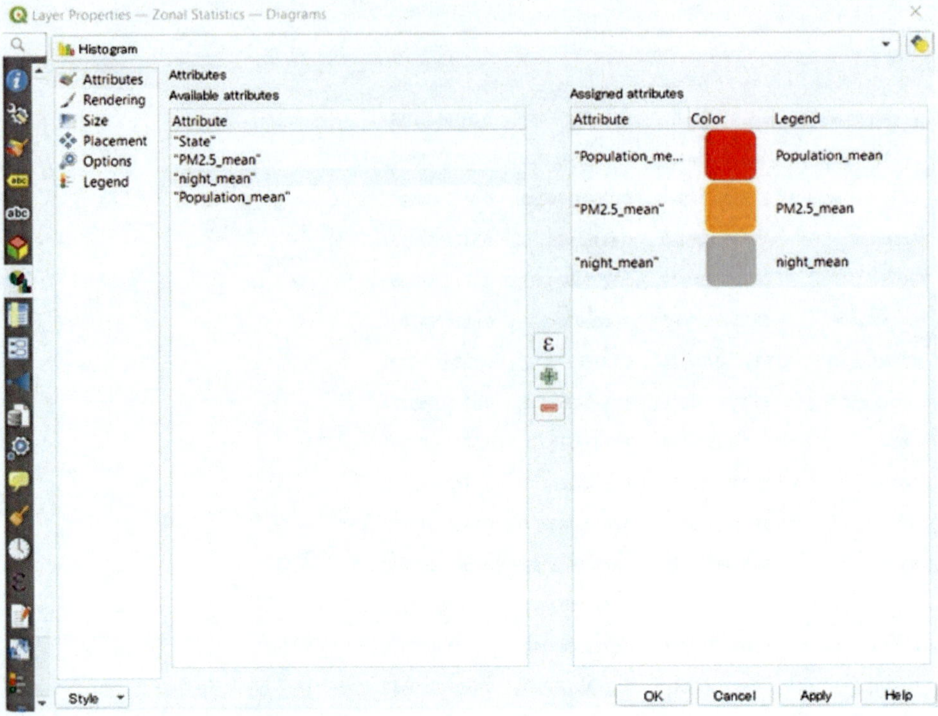

Fig. 7.33 Display the diagram

Click on "Size," select the attribute, then adjust the "Maximum value" by clicking "Find" and "Bar length," then click Apply (Fig. 7.34).

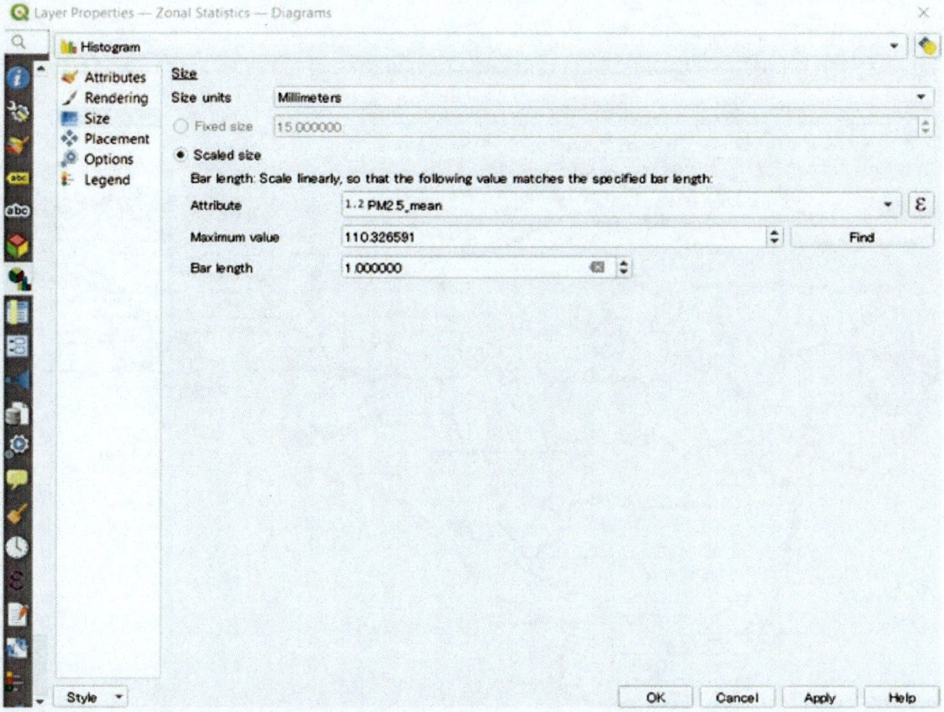

Fig. 7.34 Change the size of diagram

The result will be displayed in the main canvas of QGIS as follows. As shown the result is not good enough since the range value of each attribute is too large (Fig. 7.35). Therefore, we need to do normalization.

Fig. 7.35 Map overlaid with diagram before normalization

Normalization can be done using field calculator (Fig. 7.36). Open the attribute of layer and open the field calculator. Check on "Create a new field" and type the field name in "Output field name." Change the "Output field type" to "Decimal number (numeric)" and set the "Output field length."

In the expression window type the following expression:

"PM2.5_mean"/maximum ("PM2.5_mean")- minimum ("PM2.5_mean")
"night_mean"/maximum ("night_mean")- minimum ("night_mean")
"Population_mean"/maximum ("Population_mean")- minimum ("Population_mean")

Do it one by one for each new field. The first expression is normalization formula for new field "PM.25_norm," while the second and last expressions are normalization formula for new fields "NTL_norm" and "Population_norm," respectively.

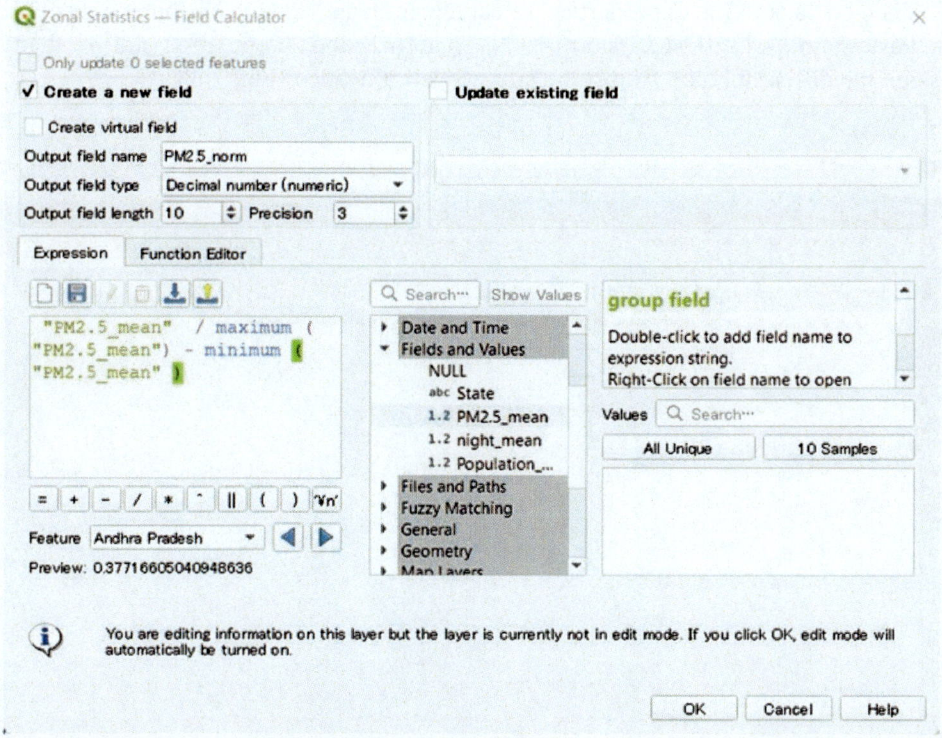

Fig. 7.36 Normalization process using Field Calculator

After finishing, the attribute table will have three new fields as shown in Fig. 7.37.

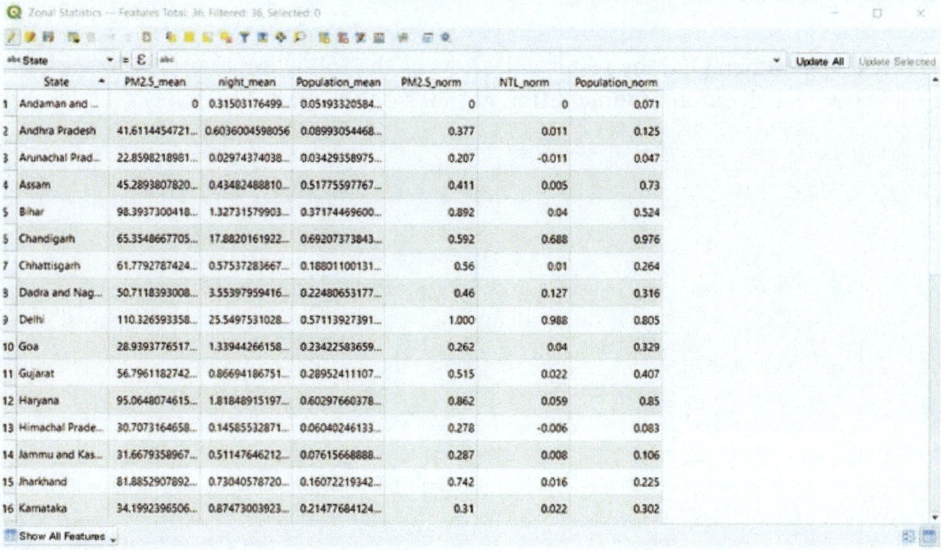

	State	PM2.5_mean	night_mean	Population_mean	PM2.5_norm	NTL_norm	Population_norm
1	Andaman and ...	0	0.31503176499...	0.05193320584...	0	0	0.071
2	Andhra Pradesh	41.6114454721...	0.6036004598056	0.08993054468...	0.377	0.011	0.125
3	Arunachal Prad...	22.8598218981...	0.02974374038...	0.03429358975...	0.207	-0.011	0.047
4	Assam	45.2893807820...	0.43482488810...	0.51775597767...	0.411	0.005	0.73
5	Bihar	98.3937300418...	1.32731579903...	0.37174469600...	0.892	0.04	0.524
6	Chandigarh	65.3548667705...	17.8820161922...	0.69207373843...	0.592	0.688	0.976
7	Chhattisgarh	61.7792787424...	0.57537283667...	0.18801100131...	0.56	0.01	0.264
8	Dadra and Nag...	50.7118393008...	3.55397959416...	0.22480653177...	0.46	0.127	0.316
9	Delhi	110.326593358...	25.5497531028...	0.57113927391...	1.000	0.988	0.805
10	Goa	28.9393776517...	1.33944266158...	0.23422583659...	0.262	0.04	0.329
11	Gujarat	56.7961182742...	0.86694186751...	0.28952411107...	0.515	0.022	0.407
12	Haryana	95.0648074615...	1.81848915197...	0.60297660378...	0.862	0.059	0.85
13	Himachal Prade...	30.7073164658...	0.14585532871...	0.06040246133...	0.278	-0.006	0.083
14	Jammu and Kas...	31.6679358967...	0.51147646212...	0.07615668888...	0.287	0.008	0.106
15	Jharkhand	81.8852907892...	0.73040578720...	0.16072219342...	0.742	0.016	0.225
16	Karnataka	34.1992396506...	0.87473003923...	0.21477684124...	0.31	0.022	0.302

Fig. 7.37 New fields in the attribute table

Then go back to Layer properties, select Diagram and add the attribute to display in Histogram. Add "PM.25_norm," "NTL_norm," and "Population_norm" then assign the different color. Adjust the size as well if needed (Fig. 7.38).

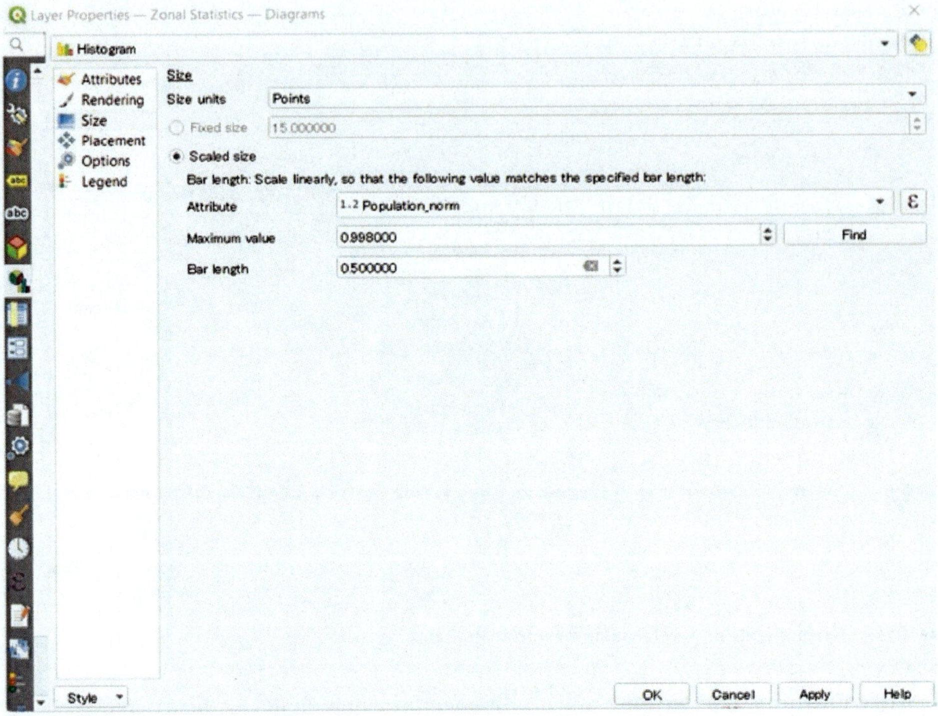

Fig. 7.38 Change the diagram data

Click Apply, then OK; the result will show as the following figure. Now we can see a better visualization of three different fields on the map (Fig. 7.39).

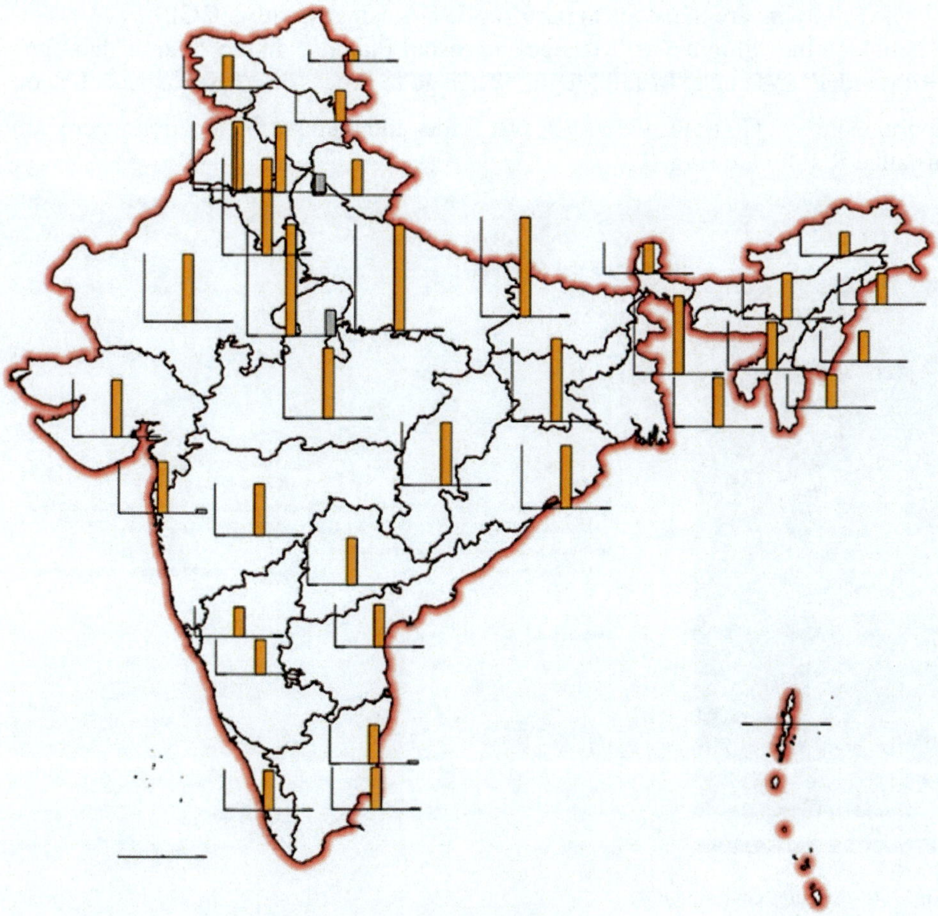

Fig. 7.39 Map overlaid with diagram after normalization

7.5.3 *Explanatory Analysis*

Exploratory spatial data analysis (ESDA) is a critical step in understanding geographic data. ESDA is the process of analyzing data to understand its characteristics and to identify patterns, relationships, and trends in the data. QGIS provides a powerful tool for conducting ESDA. To explore relationships between different layers, for example, you can use the spatial join tool to join two layers and see how they are related. Or you can analyze using simple linear regression and see how the relationship between the spatial features.

R is a popular programming language and environment for statistical computing and graphics, and now we can use R within QGIS conveniently. In QGIS, you can use the R console to perform ESDA on your geospatial data. Before you can use R within QGIS environment, you have to download and install R and RStudio first. It is available at https://posit.co/download/rstudio-desktop/

The following are the steps to perform ESDA using R within QGIS:

Click on the “Plugins” > “Manage and Install Plugins.” In the “search” tab, type “R Provider” and click “Install Plugin.” Then go to Processing Toolbox and click on option icon . Go to Providers > R > R folder and then set the location where you installed R software (Fig. 7.40).

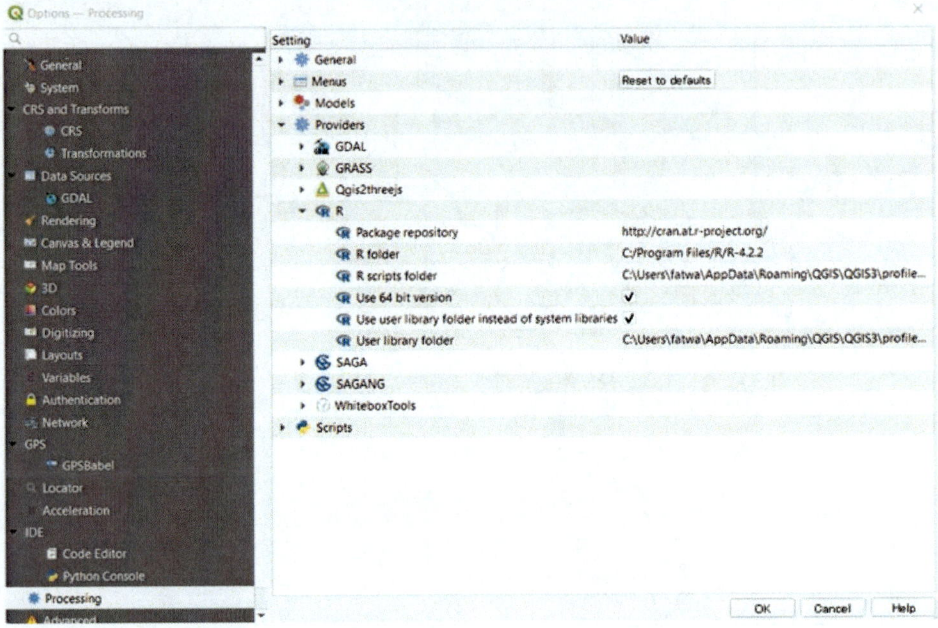

Fig. 7.40 Setting R within QGIS

Next, go back to “Plugins” > “Manage and Install Plugins,” then type “QGIS Resource Sharing” and click “Install Plugin.” Then look for small icon click on it and type “QGIS R” then select “QGIS R Collection (QGIS Repository…)” and click “Install” (Fig. 7.41).

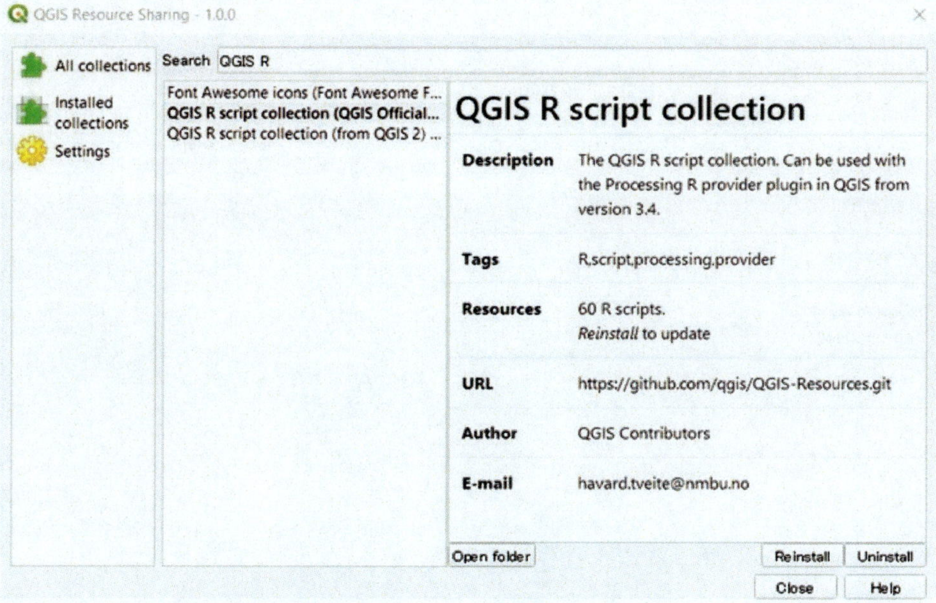

Fig. 7.41 Install QGIS R collection plugin

Finally, close and open again your QGIS, then try to execute one script under R from Processing Toolbox. For instance, double-click on "scatterplot regression," select the layer and field, and assign a title as shown in the following Figure, then click "Run" (Fig. 7.42).

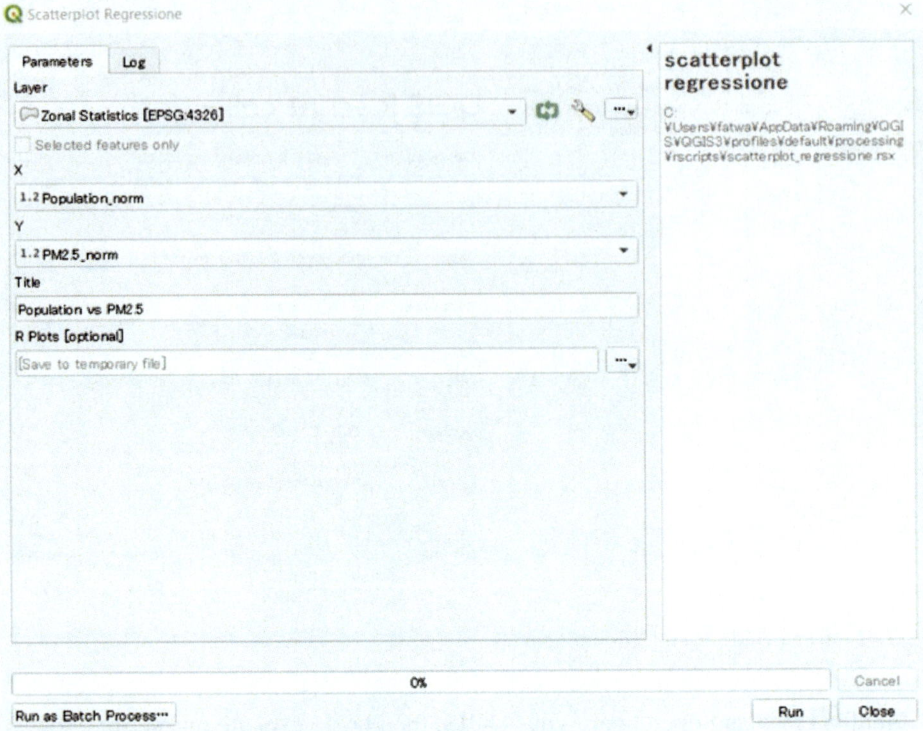

Fig. 7.42 Run the scatterplot regression

After couple seconds, the process will be finished and to open the scatterplot double-click on "R Plots" at "Results Viewer" tab as shown in Fig. 7.43.

Fig. 7.43 Result viewer

The Internet browser will open and visualize the scatterplot as Fig. 7.44. To save it, just right-click then select "Save image as".

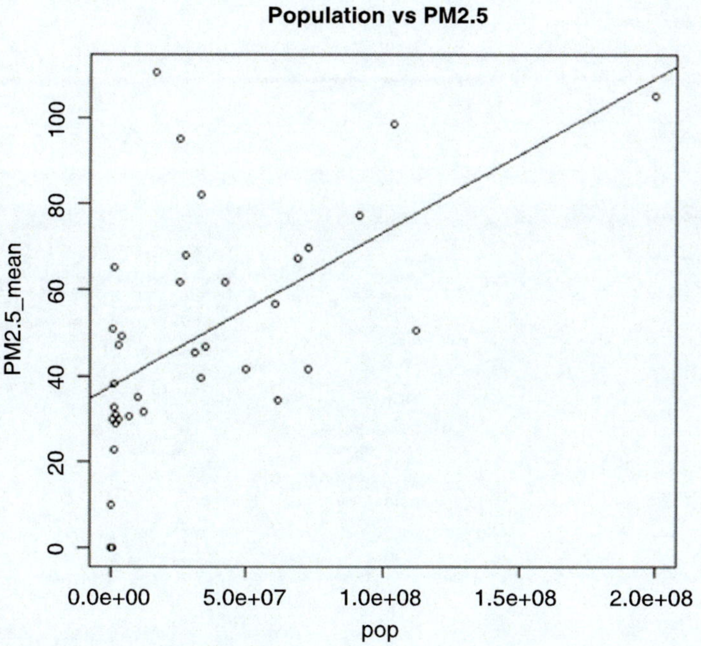

Fig. 7.44 Result of scatterplot regression

Other options are "Simple Linear Regression" where you can plot four different figures such as residual vs. fitted, normal q-q, scale-location, and Cook's distance. In this case, we will evaluate the relationship between population density and PM2.5. You need to "Create a new field" of "Area (Km2)" and calculate the area based on the polygon using "Field Calculator" and just type "$area*1e-6" as shown in Fig. 7.45. Using this field you can calculate the population density (population/area) using Field Calculator.

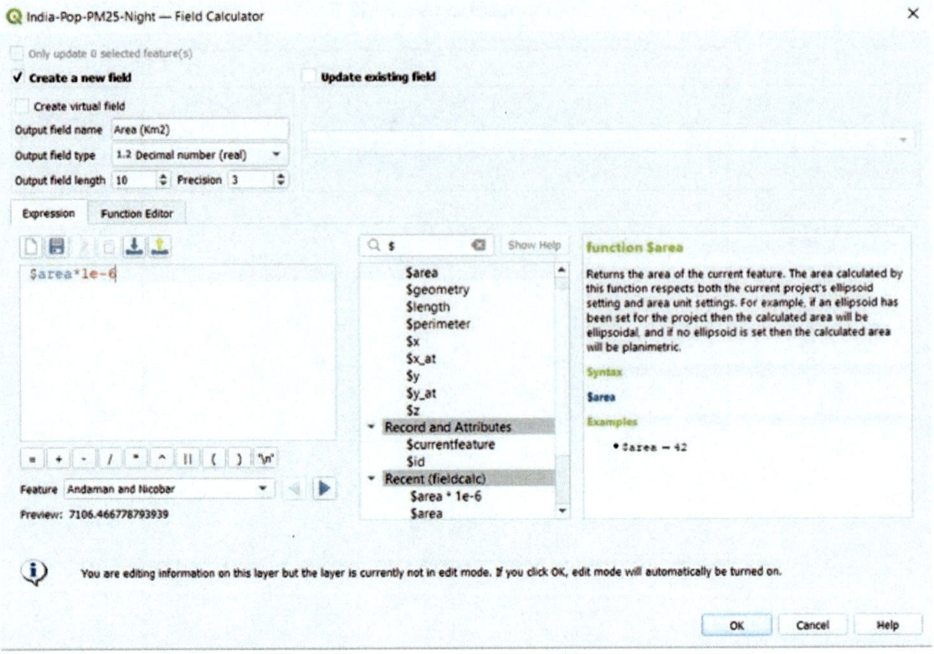

Fig. 7.45 Create a new field

After finishing, go back to Processing Toolbox and double-click on "Simple Linear Regression." Select the layer, Field1 (Pop Density) and Field2 (PM2.5) (Fig. 7.46).

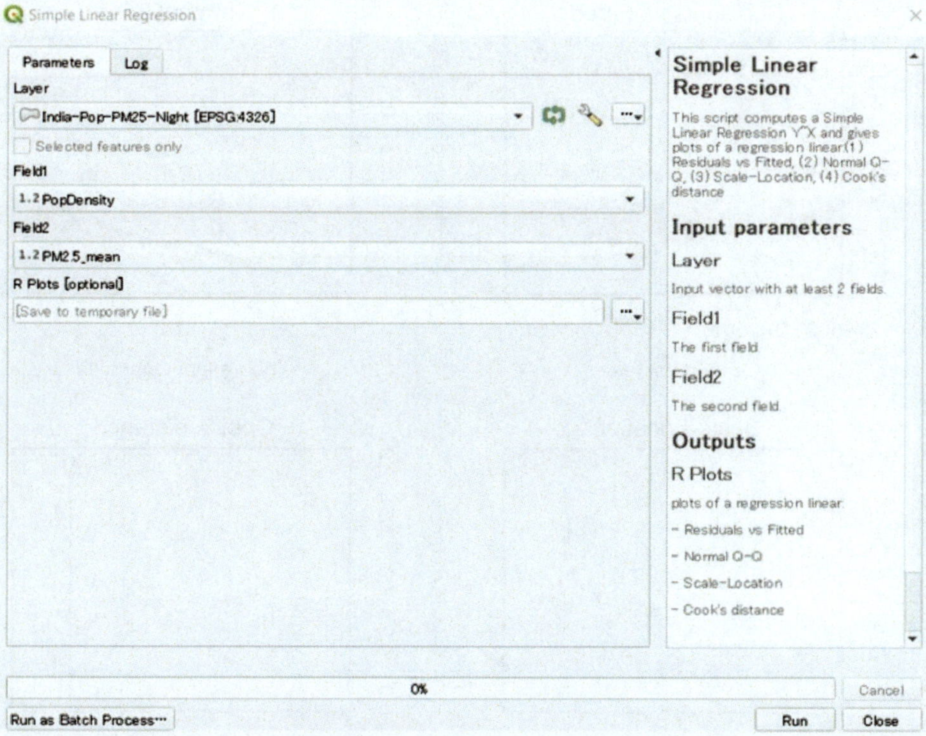

Fig. 7.46 Run the simple linear regression

After clicking Run, wait for couple seconds and the result will be available after you click on "R Plots" at "Results Viewer" tab. The result will be displayed at your Internet browser (Fig. 7.47).

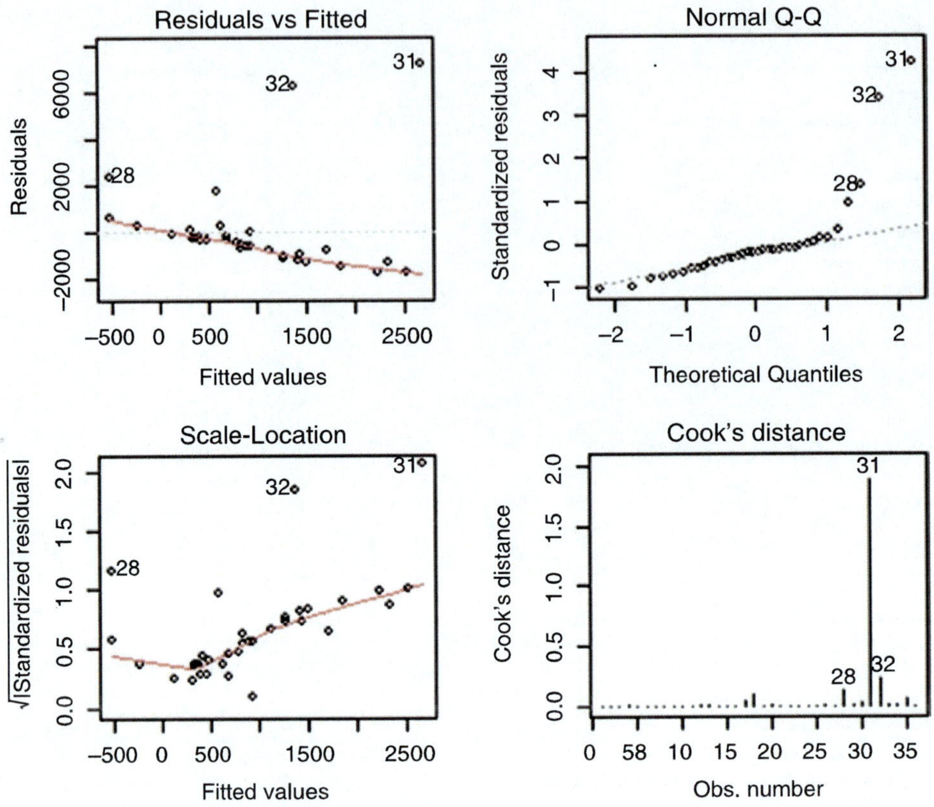

Fig. 7.47 Result of statistical analysis

In statistics, residual vs. fitted, normal q-q, scale-location, and Cook's distance are all diagnostic plots commonly used to assess the goodness-of-fit of regression models. A residual vs. fitted plot is a scatter plot of the residual values (the difference between the observed response and the predicted response) against the fitted values (the predicted response). It is used to detect patterns in the residuals, which could indicate nonlinearity, heteroscedasticity, or outliers.

While the normal quantile-quantile (Q-Q) plot is a graphical method for comparing the distribution of the residuals to a normal distribution, the residuals are plotted against theoretical quantiles from a normal distribution. If the residuals are normally distributed, the points on the plot should form a straight line.

The scale-location plot is a scatter plot of the square root of the absolute value of the standardized residuals against the fitted values. It is used to detect heteroscedasticity (non-constant variance) in the residuals. If the residuals are homoscedastic, the points on the plot should be randomly scattered around a horizontal line.

Finally, the Cook's distance is a measure of the influence of each observation on the estimated regression coefficients. It is calculated by comparing the coefficients estimated with and without each observation, and measuring the difference in the fit. Large values of Cook's distance indicate influential points that may be outliers or have a disproportionate effect on the model.

Residual vs. fitted, normal q-q, scale-location, and Cook's distance are all useful diagnostic plots for checking the assumptions of linear regression models and identifying potential problems with the data. ESDA using R within QGIS is a powerful tool for analyzing spatial data. With R scripts, you can perform a wide range of analysis tasks to better understand your data and identify patterns, relationships, and trends.

7.5.4 Create or Modify the R Script

One of the QGIS strengths is the ability to integrate with programming languages such as R, allowing for even more advanced data analysis and processing capabilities.

To create or modify an R script within QGIS, you can follow these steps:

Go to the Processing Toolbox, expand the "R Scripts" category to see the available R scripts.

If you want to modify an existing R script, right-click on the script and select "Edit Script" (Fig. 7.48). This will open the R script editor, where you can make your changes. In this case we will modify "Multiple Regression" from using seven variables to only using three variables (Fig. 7.49).

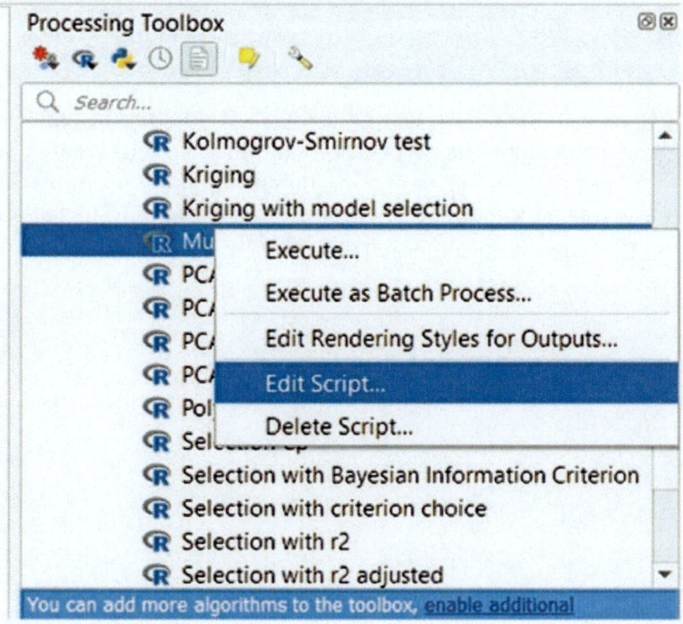

Fig. 7.48 Edit the R script

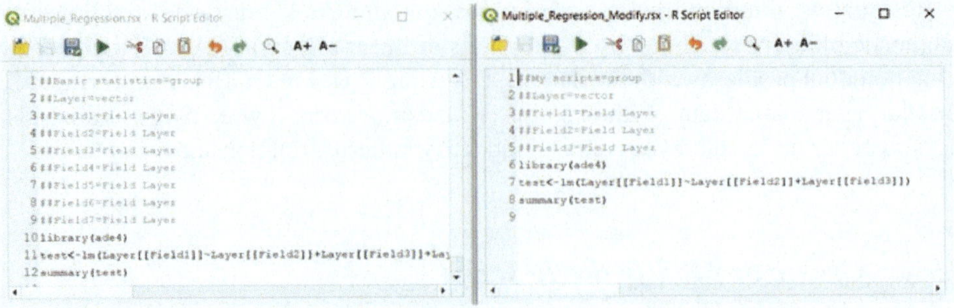

Fig. 7.49 Edit the number of input variables

If you want to create a new R script, click on the "R" category and select "Create New R Script." This will open the R script editor with a template script (Fig. 7.50).

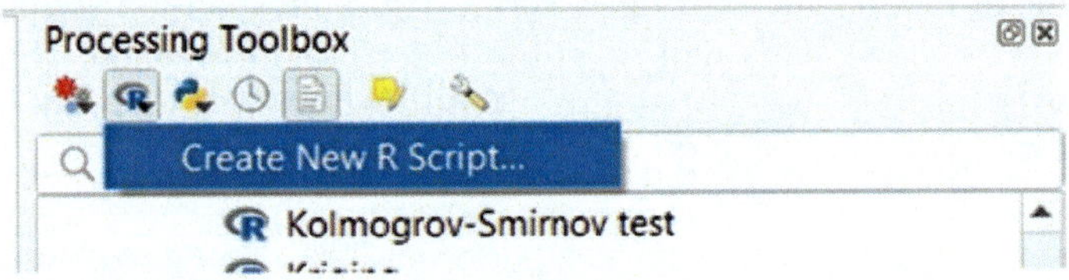

Fig. 7.50 Create new R script

Write your R code in the editor, using the QGIS-R integration functions to interact with the QGIS project and its layers. For example, we will create "Sample random points" function. The code is available at https://north-road.github.io/qgis-processing-r/ select "Examples > Example with vector output." Copy and paste the script in the R script editor window (Fig. 7.51).

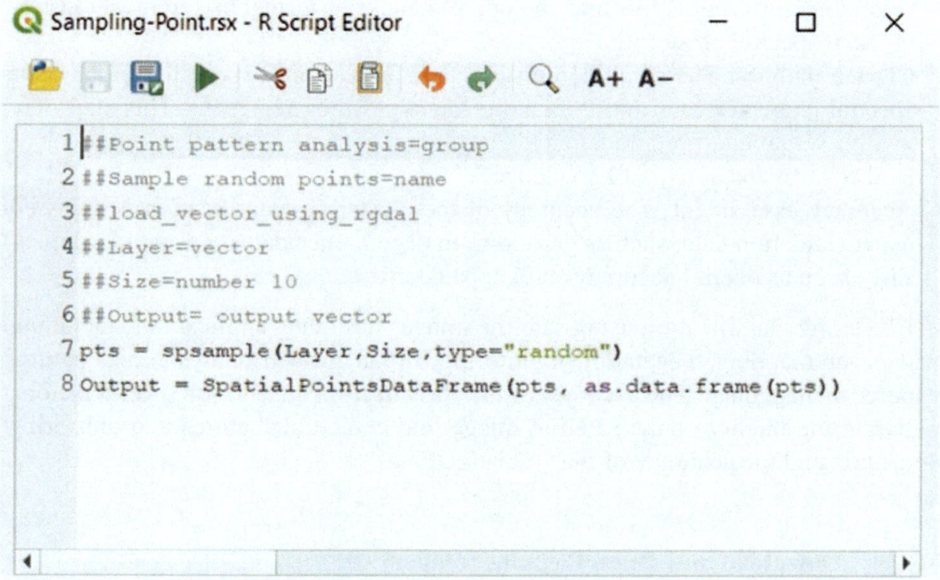

Fig. 7.51 Write the new script

Once you have finished writing your script, click on the "Save" button to save the script to a file.

You can now run the script by clicking on the "Run" button in the R script editor.

7.6 Land Use Land Cover Analysis

7.6.1 Clustering

Clustering classification using satellite images is a process of grouping or categorizing pixels in satellite images into distinct clusters or classes based on their similarity in terms of spectral characteristics. The goal is to identify areas of the image that exhibit similar reflectance or radiance properties and are likely to represent similar land cover types.

The process typically involves the following steps:

- Image preprocessing: The satellite image is preprocessed to remove noise, atmospheric effects, and other artifacts that could affect the accuracy of the clustering algorithm.
- Feature extraction: The spectral bands or indices that are most relevant for identifying land cover types are selected and extracted as features from the satellite image.
- Clustering algorithm: A clustering algorithm is applied to the extracted features to group pixels with similar spectral characteristics into distinct clusters.

Common clustering algorithms include k-means clustering, hierarchical clustering, and spectral clustering.
- Cluster analysis: The resulting clusters are analyzed and labeled based on their spectral characteristics and the land cover types they represent. This step may involve visual interpretation, field data validation, or use of ancillary data such as land cover maps or expert knowledge.
- Accuracy assessment: The accuracy of the clustering classification is assessed using validation data, such as ground truth or reference data, and statistical measures such as overall accuracy and kappa coefficient.

Clustering classification using satellite images has many applications, including land cover mapping, vegetation monitoring, urban growth analysis, and natural resource management. The accuracy of the classification depends on several factors, including the quality of the satellite image, the choice of features and clustering algorithm, and the accuracy of the validation data.

7.6.1.1 Download and Open Pleiades Imagery

Please follow these steps to access the sample data of Pleiades image:

1. Go to https://www.intelligence-airbusds.com/imagery/sample-imagery/pleiades-neo-hd15-marseille-france/ and fill out the form.
2. After a couple of minutes, you will receive an email with a URL to download the sample data.
3. Download the data and extract it.
4. Look for the file named "IMG_PNEO3_202111071029298_PMS-FS_ORT_PWOI_000006529_1_1_F_1_NED_R1C1.tif" inside the main folder of "Marseille_Ortho_Pan-Sharpened_DIMAP_Geotiff_12bits_Reflectance."
5. Open QGIS and load the imagery.
6. Change the RGB composite to Green-Red-Blue.
7. Zoom in to the bottom-right corner of the image until you can see the Orange Vélodrome, as shown in Fig. 7.52.

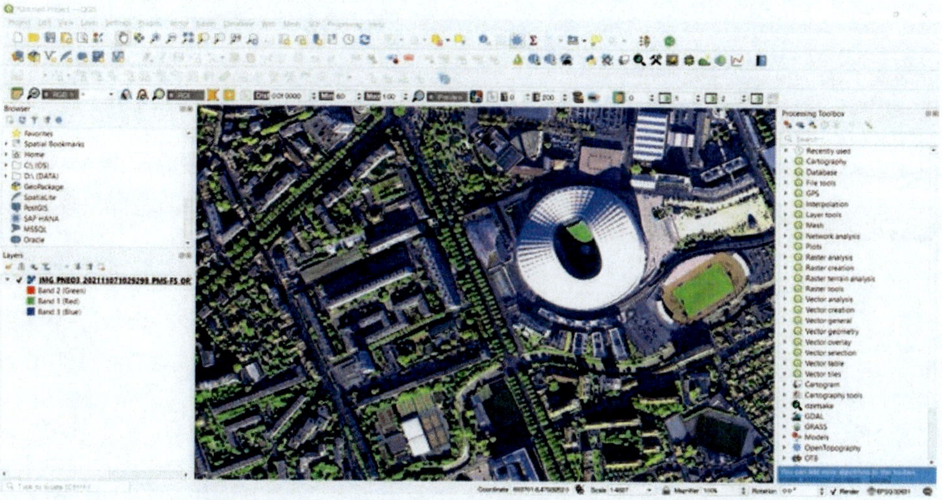

Fig. 7.52 Orange Velodrome

7.6.1.2 Create Box and Clip the Image

We need to create small dataset to minimize the computational time. The "box" layer will be utilized to clip the image. To create these layers, navigate to Layer > Create Layer > New shapefile layer, select "Polygon" as the Geometry type, enter "box" as the File Name, and click OK (Fig. 7.53).

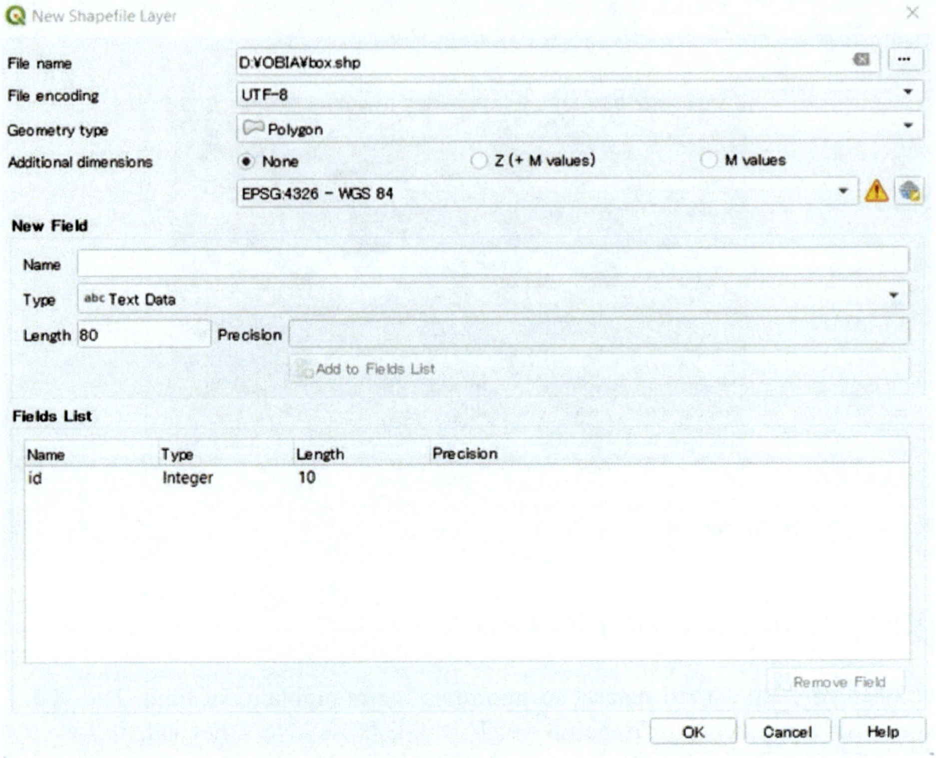

Fig. 7.53 Create box to clip the raster data

Then, locate the icon  and click on it. This will activate the layer. To begin creating the box, start by click on  and clicking on the top-right corner, then move to the top-left, bottom-left, and bottom-right corners, before finally right-clicking to finish. Next, type "01" in the window that appears and click OK to assign an ID to the box. Once you have done this, click on the icon again and select "Save" to save your work. The final result should look similar to Fig. 7.54.

Fig. 7.54 Box layers

To clip the raster data, follow these steps: Go to Raster > Extraction > Clip Raster by Extent. Select the input layer and choose "Calculate from layer > box" as the Clipping extent. Finally, click on the Run button to execute the command (Fig. 7.55).

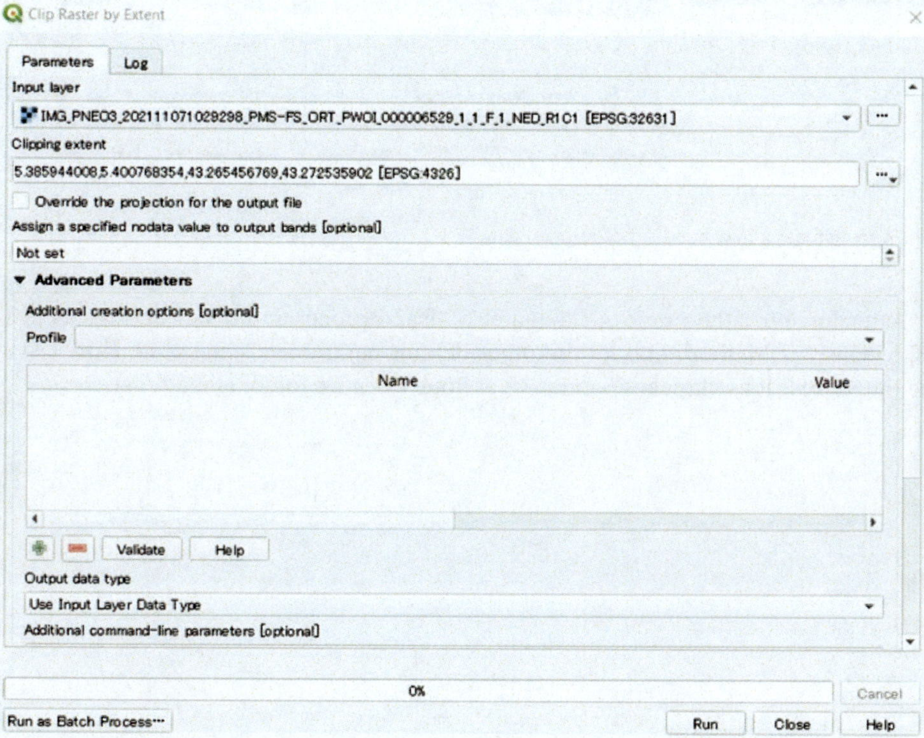

Fig. 7.55 Clip raster by extent

7.6.1.3 Clustering Using K-Means

Next, navigate to icon  then go to "Band set," in the "Multiband image list" select the clipped dataset. In the "Wavelength quick settings" select "Band order" and check on "Create virtual raster" and "Create raster of (stack bands)" then click Run (Fig. 7.56).

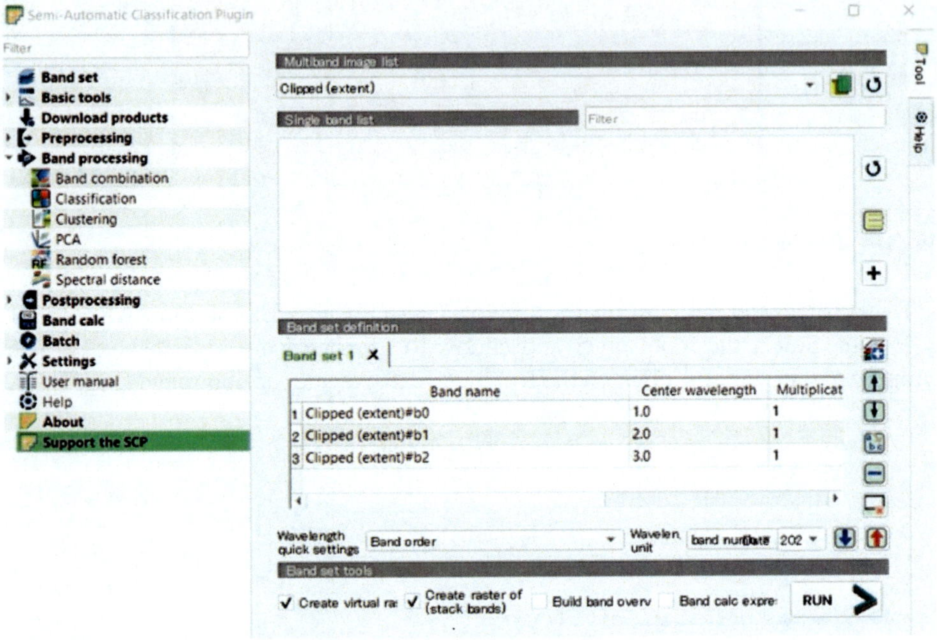

Fig. 7.56 Create a new band set of clipped data

After finishing the process, navigate to "Band processing > Clustering," select "K-Means" as Method, and let the other default parameters, then click Run. Give the output result as "kmeans" and save at your working folder (Fig. 7.57).

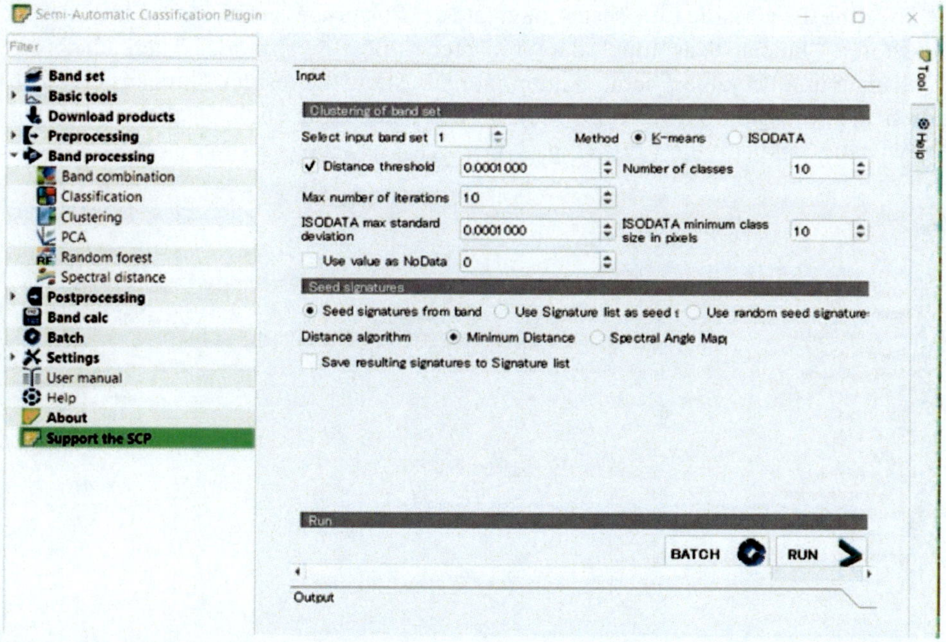

Fig. 7.57 K-Means clustering

After couple minutes, the result will be displayed in the main canvas of QGIS automatically. While in the SCP, the Class ID, Signature, and Distance values of each class will appear in window as Fig. 7.58.

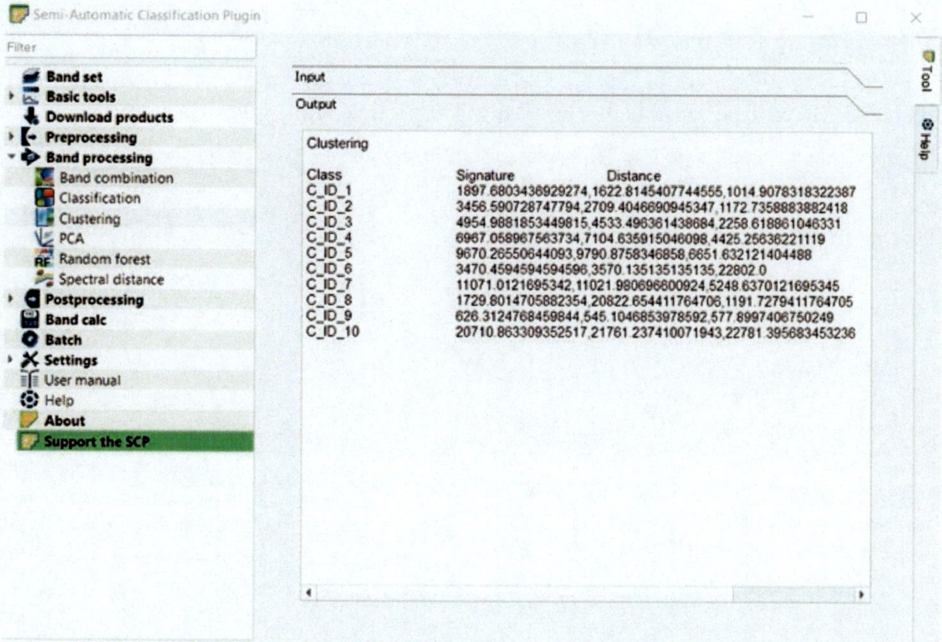

Fig. 7.58 K-Means clustering result

To refine the classification result, navigate to "Postprocessing > Reclassification." Select the "kmeans" as input data for reclassification process. Then click on "Calculate unique values" icon to automatically show the result of kmeans classification in the table below. To do reclassification, double-click on the row under "New value" and edit the value (Fig. 7.59).

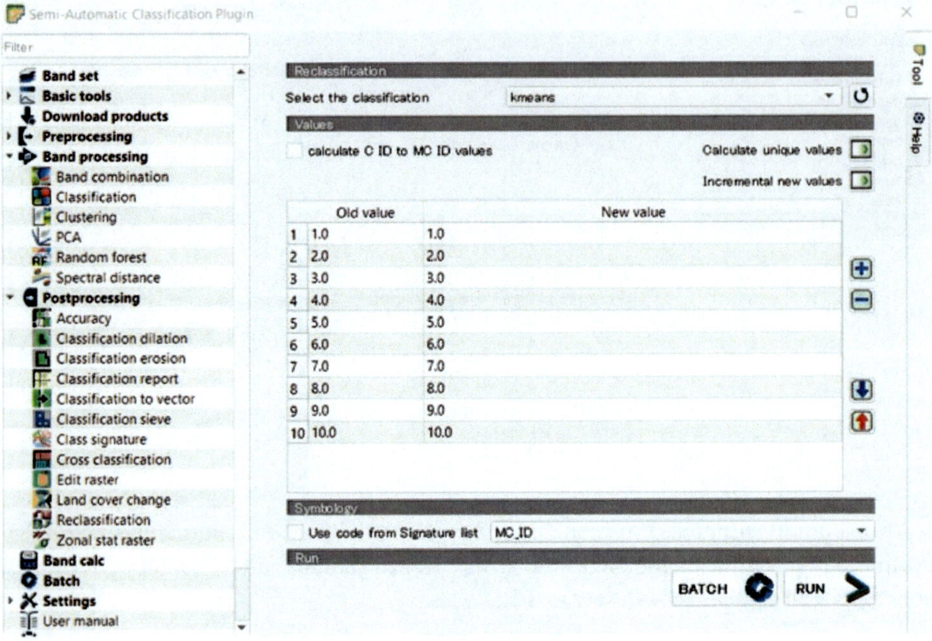

Fig. 7.59 Reclass the K-Means result

The result will be similar as Fig. 7.60. Then click Run, give it a name such as "kmeans-reclass."

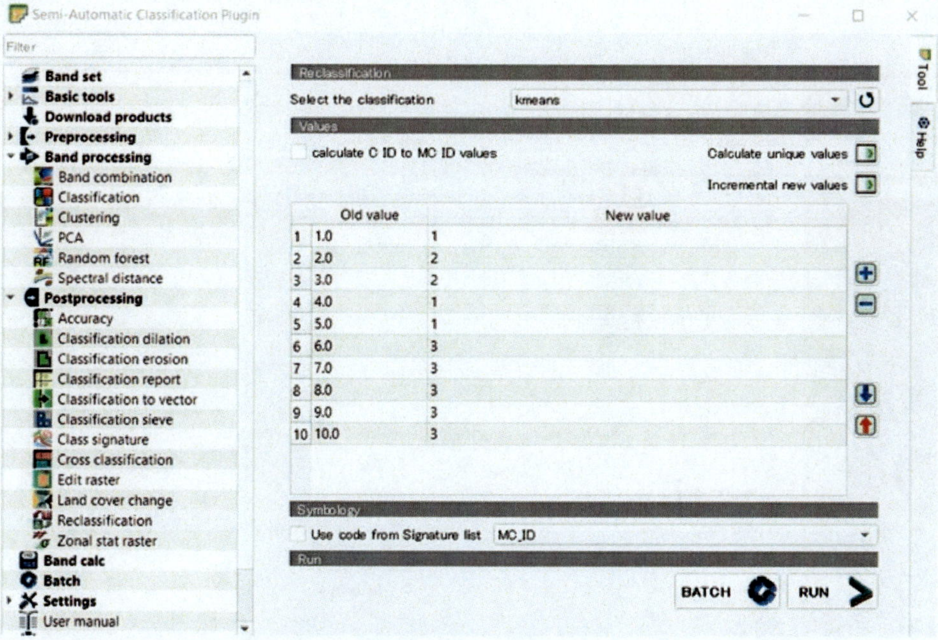

Fig. 7.60 Table of reclass result

The result will be displayed in "Singleband gray." To change it, double-click on the "kmeans-reclass" layer and change the symbology to "Paletted/Unique values," then click OK (Fig. 7.61).

Fig. 7.61 Change the visualization of K-Means reclass

Final result of the clustering process is as shown in Fig. 7.62.

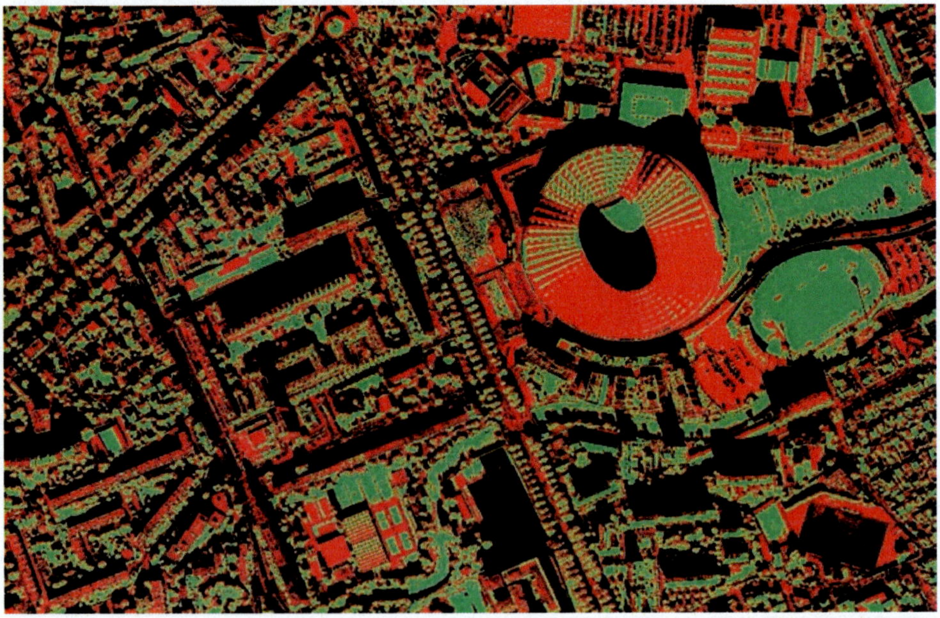

Fig. 7.62 Result of clustering classification using K-Means algorithm

7.6.1.4 Classification Report

We can also calculate the area of each class. Navigate to "Postprocessing > Classification report." Select the classification of "kmeans-reclass" then click Run (Fig. 7.63). If the "kmeans-reclass" does not appear, click on the refresh icon. Give it name, for example, "kmeans-report," it will be saved as .CSV file.

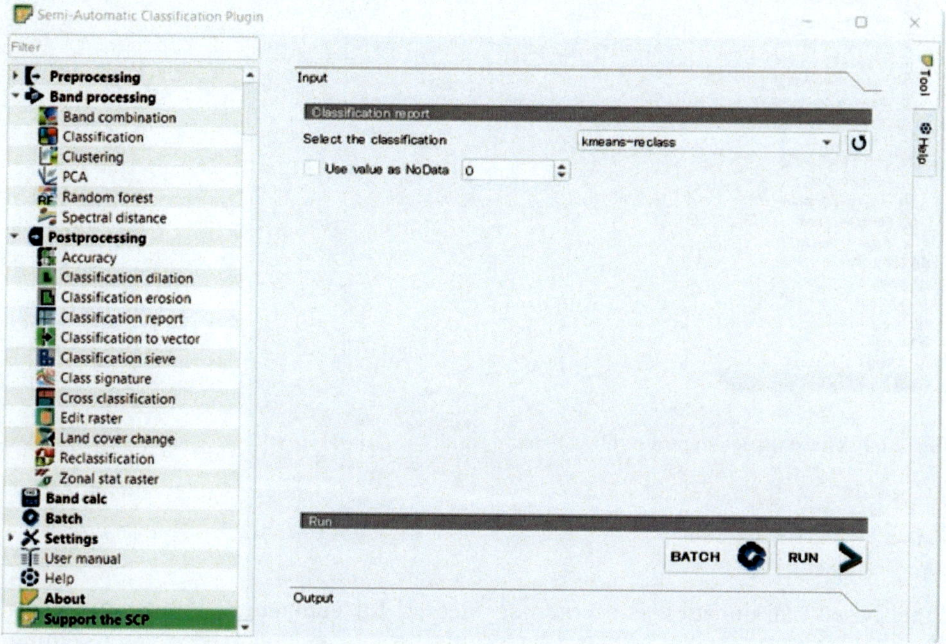

Fig. 7.63 Calculate the classification result

The result will also be displayed as Fig. 7.64.

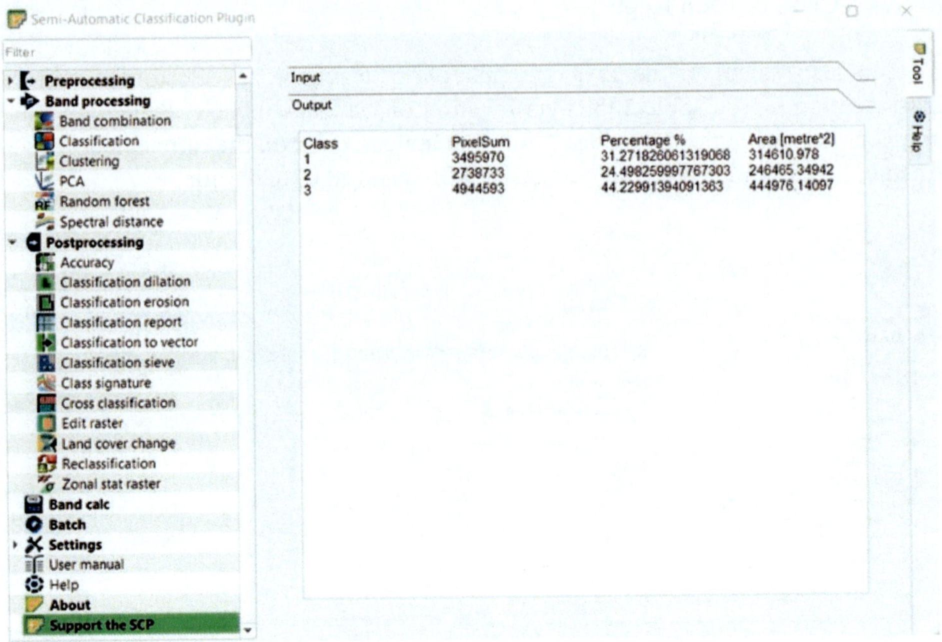

Fig. 7.64 The statistical report of classification result

7.6.2 *Pixel-Based*

Pixel-based classification is a common method for analyzing satellite images to identify different features and land cover types in a given area. It involves assigning each pixel in an image to a specific land cover or feature class based on its spectral characteristics.

Satellite images are composed of a series of pixels, each of which represents a specific location on the Earth's surface. These pixels are typically classified based on the reflectance values of different wavelengths of light that are captured by the satellite's sensors. The most common way to do this is by using a technique called supervised classification, which involves training a computer algorithm to recognize different land cover classes based on a set of training data.

To perform a pixel-based classification using satellite imagery, the following steps are typically followed:

- Preprocessing: The satellite image is first preprocessed to remove any distortions or artifacts that may be presented in the data.
- Selection of training data: A set of representative pixels from the image is selected, each of which is labeled with a specific land cover class. These pixels are used to train the classification algorithm.
- Feature extraction: The spectral characteristics of each pixel in the image are extracted, usually in the form of a set of numerical values representing the reflectance values at different wavelengths.

- Training the algorithm: A classification algorithm, such as maximum likelihood or support vector machines, is trained using the labeled training data and the extracted features.
- Classification: The trained algorithm is then applied to the entire satellite image, with each pixel being assigned to a specific land cover or feature class based on its spectral characteristics and the classification algorithm.
- Postprocessing: The resulting classified image may be further processed to remove noise, fill in gaps, or refine the classification boundaries.

Pixel-based classification is a powerful tool for analyzing satellite images and can be used to monitor changes in land cover over time, identify areas at risk of natural disasters, and support a wide range of environmental and land management applications.

7.6.2.1 Creating Training Data

To create training data, follow these steps: navigate to Layer > Create Layer > New shapefile layer. Next, select "Polygon" as the Geometry type, enter "training" as the File Name and do not forget to create a new folder if needed, and click OK. Please refer to the Table 7.1 for guidance on creating the training polygon and assigning class IDs.

Table 7.1 Class and ID for pixel-based classification

Class	id
Built up	1
Tree	2
Building	3
Vegetation	4
River	5

To start creating the training polygon, click on the  icon and then proceed to click on the  icon. Once you have completed the polygon, right-click and enter the corresponding ID. Please refer to the "Built up" (1), "Waste" (2), "Building" (3), "Vegetation" (4), and "River" (5) as examples below for a clear demonstration (Fig. 7.65).

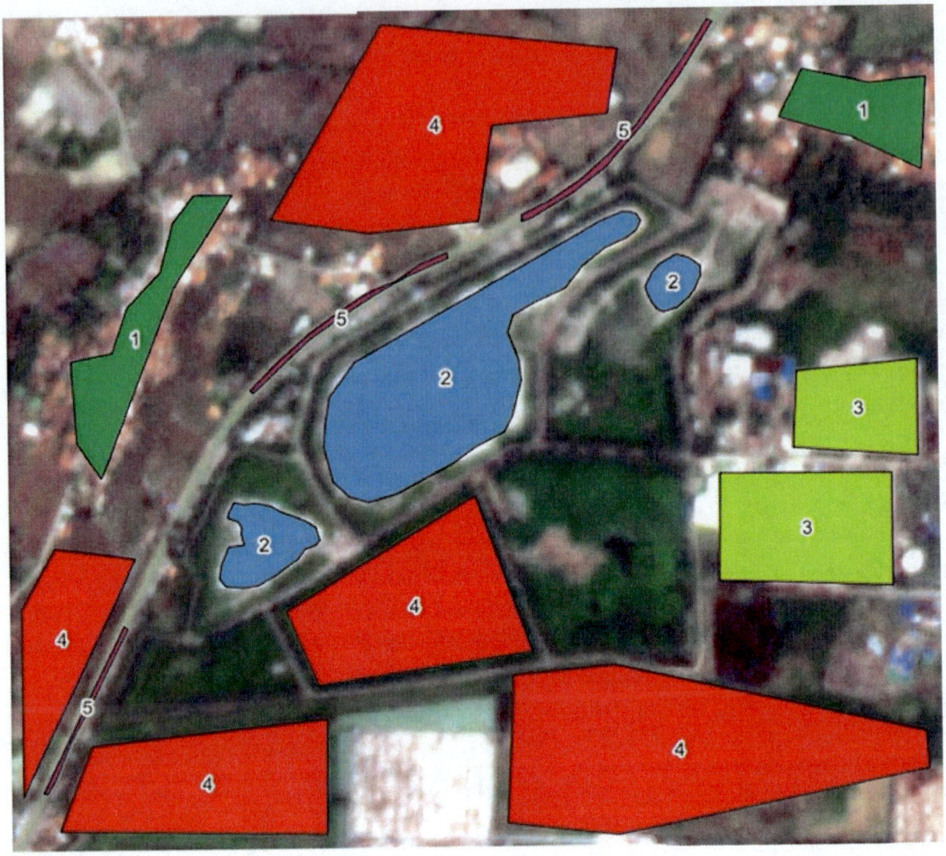

Fig. 7.65 Training polygons

You have to make sure that the number of training polygon is evenly distributed. Do not forget to click on "Save Layer Edits" button  and then click on the  icon to finish and save your training layer.

At first, your training polygon layer will be in "Single Symbol" that is why the layer is only shown in a single color. To change it, double-click on your "training" layer and select "Symbology." Change from "Single Symbol" to "Categorized" in the first drop-down menu on the right side. On the "Value," select "id" then click "Classify" button at the bottom-left, then click "OK."

7.6.2.2 Install Dzetsaka Plugin

To do pixel-based classification, we need to install the dzetsaka plugin first. It is a plugin for semi-automatic classification with three different classifiers available: Gaussian Mixture Model, Random Forest, and SVM. However, before we use it we have to make sure the scikit-learn library has been installed in our system first.

To install the plugin go to Plugin-Manage and Install Plugin then type "dzetsaka" in the search window. Check on the left small-box next to "dzetsaka : Classification tool" then click "Install Plugin." Then look for this small icon  and click. The dzetsaka user interface will appear on the bottom-left of your QGIS window (Fig. 7.66).

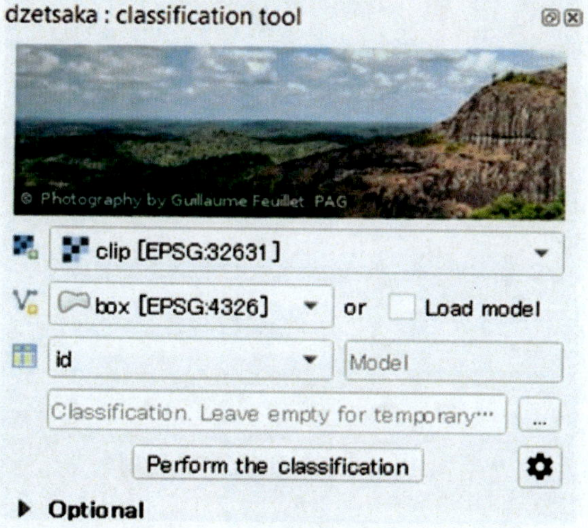

Fig. 7.66 The dzetsaka plugin user interface

7.6.2.3 Select Classifier and Perform Classification

To do classification, change the input raster layer to "clip" and shapefile training data to "training" and click on setting icon to change the classifiers algorithm. Change from Gaussian Mixture Model to Random Forest and click on X button. Then click on "Perform the classification" and wait until it finishes.

The classification process will take time; it depends on some factors such as your computer system specifications, size of satellite data, and the classifier algorithm. The classification could take a lot of time, so be patient, even if the progress bar is stuck at 83%.

When using the dzetsaka plugin, you must make sure your training data satisfies the following conditions:

1. Using only integer value in your id column.
2. Raster data and shapefile training data have the same projection system.

The result will be appeared as greyscale at first, then you need to do some adjustment using symbology, double-click on the classification result layer and select Symbology, change Render type to "Paletted/Unique values." The final result will be shown as Fig. 7.67, as we can see the result is not free from the *salt-and-pepper* issue.

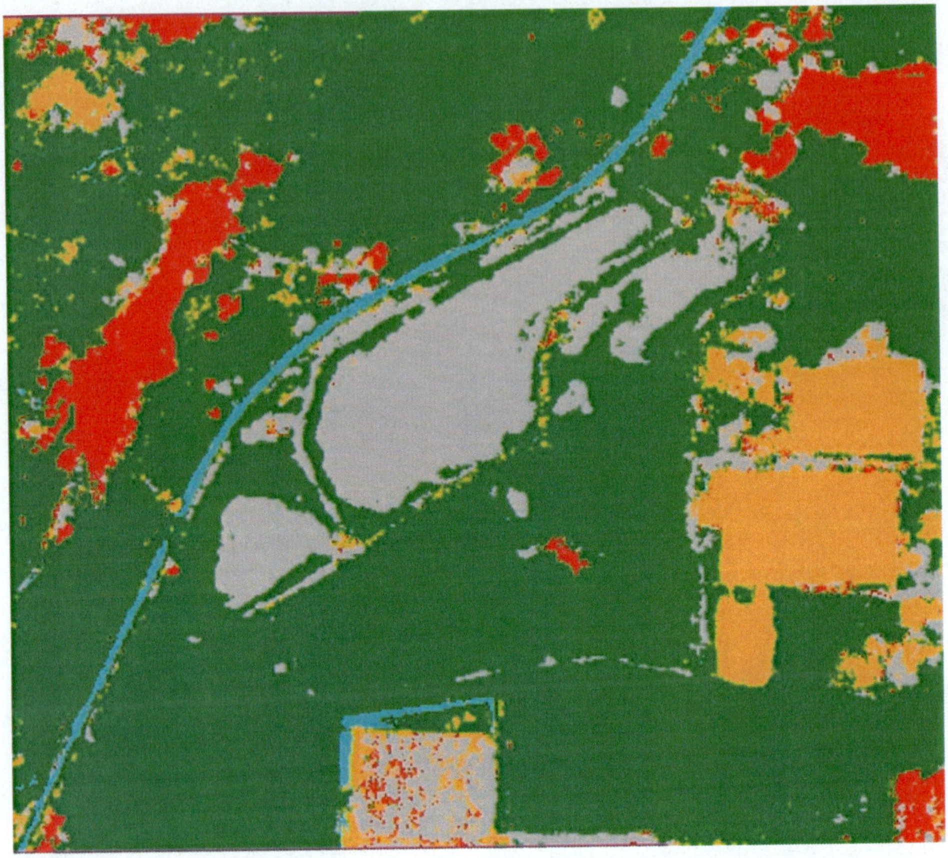

Fig. 7.67 Result of pixel-based classification using Random Forest algorithm

7.6.2.4 Advantages and Disadvantages of Pixel-Based

Pixel-based classification allows for a high level of spatial detail since each pixel is individually classified. This enables the detection of small-scale land use features and subtle changes in land cover within an image. Pixel-based methods are also flexible and can accommodate a wide range of land use classes. They can be applied to different types of remote sensing data, including multispectral and hyperspectral imagery, which makes them suitable for various applications.

Furthermore, pixel-based classification methods are often considered objective, because they rely solely on the spectral characteristics of individual pixels. They do not require prior knowledge or assumptions about the spatial arrangement or shape of land use features.

However, some disadvantages of pixel-based also need to be considered. In many cases, pixels in remotely sensed images may represent mixed land cover types. For example, a pixel may contain a combination of vegetation, buildings, and roads. Pixel-based methods struggle to accurately classify such mixed pixels, leading to misclassifications or ambiguous results.

Spectral variability within land use classes can pose challenges for pixel-based classification. Different factors like illumination, atmospheric conditions, and phenological variations can result in variations in spectral signatures of the same land use class. This variability can reduce classification accuracy. Furthermore, pixel-based methods often overlook spatial context information, as they treat each pixel independently. This can result in spatially inconsistent classifications, where neighboring pixels with similar land cover are assigned different classes.

Pixel-based methods are sensitive to the scale of analysis. The choice of pixel size or spatial resolution can influence classification results. Smaller pixels may capture more detailed information but can be susceptible to noise, while larger pixels may oversimplify complex land use patterns.

Finally, classifying individual pixels in large-scale images can be computationally intensive and time-consuming. The requirement for extensive training data and manual labeling can also add to the overall resource requirements. Land use classification from satellite images often requires skilled interpretation and expertise in image processing and remote sensing techniques. Interpreting complex scenes and distinguishing between land cover classes accurately can be challenging, requiring trained personnel or specialized algorithms.

7.6.3 Object-Based

Object-based classification is a method of classifying land cover or land use features in satellite imagery using image segmentation algorithms. It is a technique that is commonly used in remote sensing and GIS applications, particularly in the analysis of high-resolution satellite images.

The process of object-based classification involves several steps:

- Image Preprocessing: The satellite image is preprocessed to enhance the features of interest and to remove any noise or distortions.
- Image Segmentation: The image is then segmented into a series of objects or segments based on their spectral properties, texture, shape, and size.
- Object-Based Feature Extraction: Features of interest, such as land cover classes, are then extracted from the objects using a set of rules or algorithms that take into account their spectral, spatial, and contextual properties.
- Classification: Finally, the extracted features are classified into different land cover classes using statistical or machine learning algorithms.

Object-based classification has several advantages over traditional pixel-based classification methods. It allows for the inclusion of contextual information in the classification process, resulting in more accurate and reliable results. It also reduces the effect of mixed pixels, which can lead to misclassification of land cover types. Additionally, object-based classification can handle a wide range of spatial scales, making it suitable for analyzing high-resolution satellite images.

Object-based classification is a powerful tool for analyzing satellite imagery and extracting land cover information. It offers a more accurate and reliable approach to classify land cover types than traditional pixel-based classification methods.

7.6.3.1 Download and Install Orfeo Toolbox

To utilize object-based classification in QGIS, you must first download and install Orfeo Toolbox from https://www.orfeo-toolbox.org/download/ Next, navigate to the Plugin menu and select "Manage and Install Plugins." Click on "Settings," followed by "Add," and enter "Orfeo Plugin" in the Name field. Finally, paste the URL https://www.orfeo-toolbox.org/qgis/plugins.xml and click OK (Fig. 7.68).

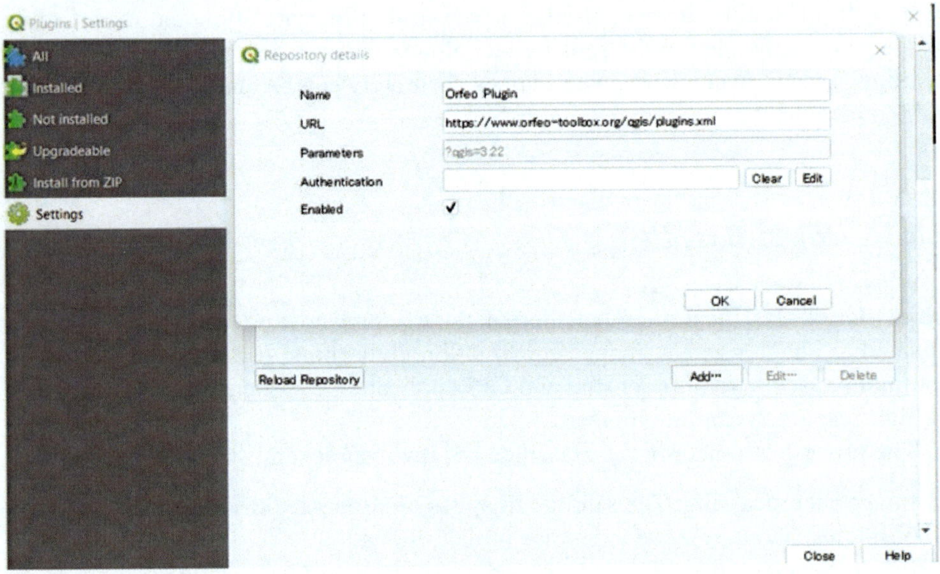

Fig. 7.68 Setting Orfeo Toolbox within QGIS

Next, navigate to the "All" section, then type "Orfeo" in the search bar and select the corresponding checkbox. Finally, click on "Close" to finish (Fig. 7.69).

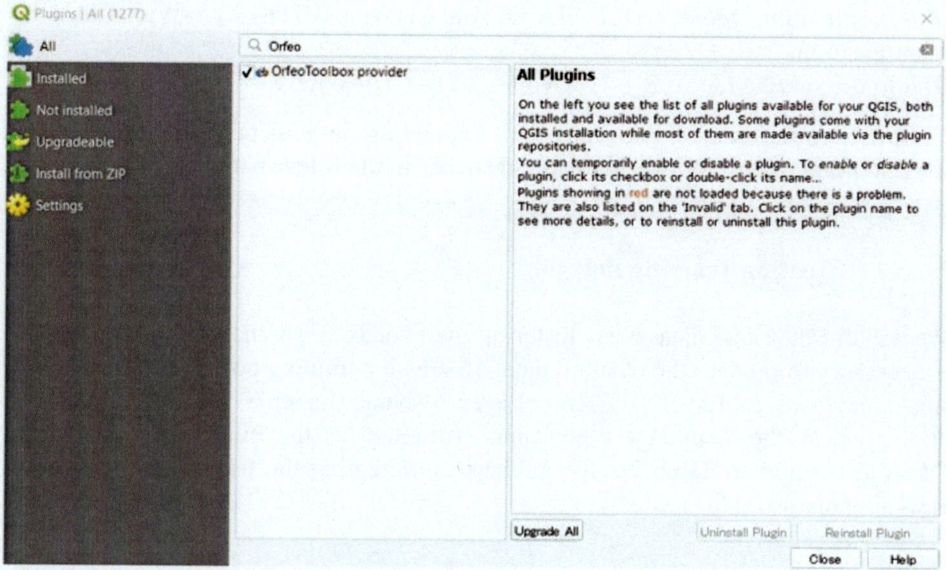

Fig. 7.69 Activate the Orfeo Toolbox plugin

Then, extract the OTB file you have downloaded, which is named "OTB-8.1.1-Win64.zip." Then, navigate to Settings > Options > Providers > OTB (Fig. 7.70), and proceed by copying and pasting the corresponding path into the following two fields:

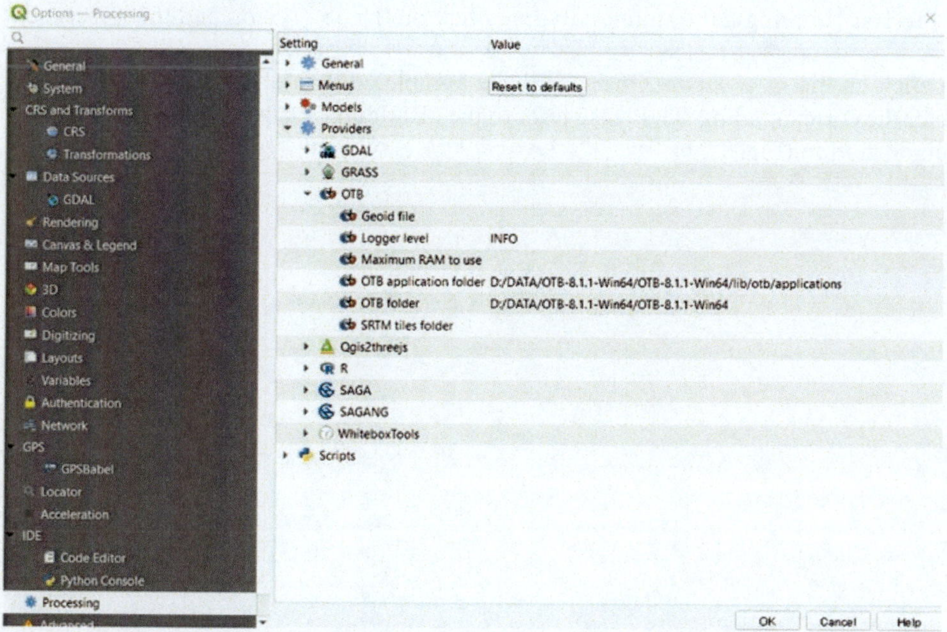

Fig. 7.70 Setting the path of OTB plugin

OTB application folder: D:\DATA\OTB-8.1.1-Win64\OTB-8.1.1-Win64\lib\otb\applications.

OTB folder: D:/DATA/OTB-8.1.1-Win64/OTB-8.1.1-Win64.

Note that the location path may vary depending on your personal preferences. It's important to ensure that there are no spaces in the folder path.

7.6.3.2 Creating Training Polygon

We will use the same dataset as clustering method. To begin the training process, it is necessary to prepare the training data. To create a training polygon, follow these steps: navigate to Layer > Create Layer > New shapefile layer. Next, select "Polygon" as the Geometry type, enter "training" as the File Name, and click OK. Please refer to Table 7.2 for guidance on creating the training polygon and assigning class IDs.

Table 7.2 Class and ID for OBIA classification

Class	id
Building	1
Tree	2
Field	3
Road	4
Shadow	5

To start creating the training polygon, click on the  icon and then proceed to click on the  icon. Once you have completed the polygon, right-click and enter the corresponding ID. Please refer to the "Building" example below for a clear demonstration (Fig. 7.71).

Fig. 7.71 Creating training polygon

After creating all the training data, we can display it as Fig. 7.72.

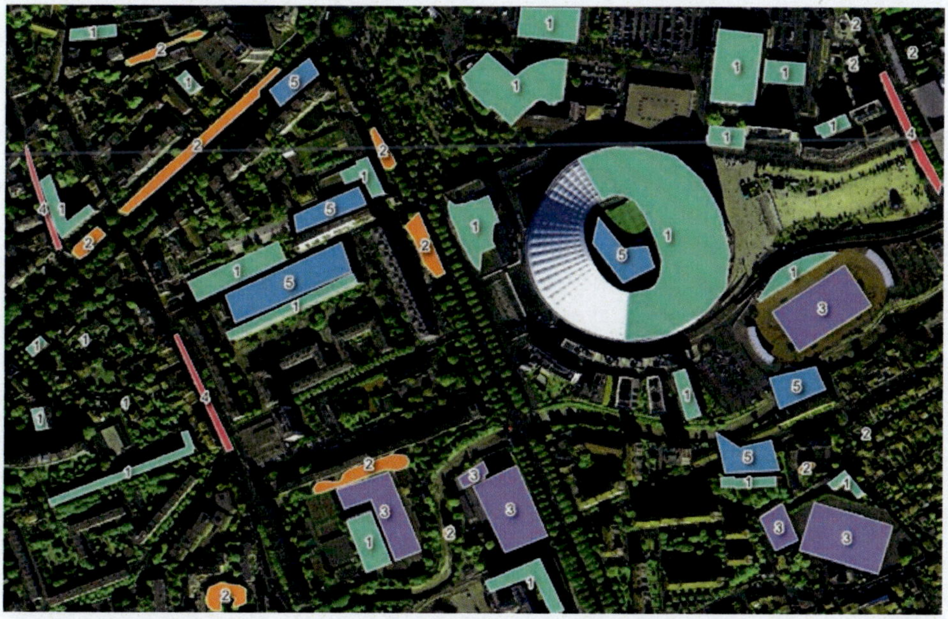

Fig. 7.72 All training polygon

7.6.3.3 Object-Based Image Analysis (OBIA)

To perform OBIA using Orfeo Toolbox, follow these steps:

1. Open the Processing Toolbox of QGIS.
2. Type "segmentation" in the search bar and double-click on "Segmentation" under "OTB > Segmentation" (Fig. 7.73).
3. Once the segmentation is completed, proceed with the following steps: Zonal Statistic, Fix geometries, Join Attribute, TrainVectorClassifier, and VectorClassifier.

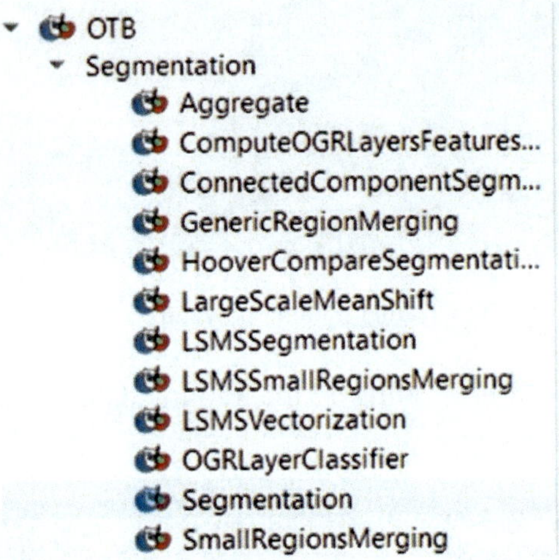

Fig. 7.73 Segmentation feature of OTB

Leave the parameters at their default settings, then scroll down and select "8-neighbor connectivity [optional]" under "Output vector file." Click on the drop-down menu and choose "Save to file" Type in a preferred file name with the file extension (e.g., "segmentation.shp") and click on "Run."

After a few minutes, the process will be completed. To add the "segmentation.shp" file, go to Layer > Add Layer > Add Vector Layer.

The next step is to calculate the Zonal Statistics. Open the Processing Toolbox and search for "ZonalStatistic." Under OTB > Image Manipulation, select "ZonalStatistics."

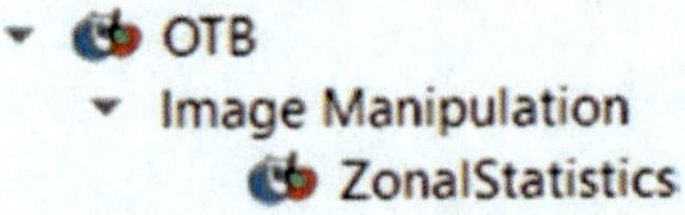

Select the clipped image in the "Input Image" field, choose the "segmentation.shp" layer in the "Input Vector Data" field, and **ensure that you include the file extension ".shp"** when specifying the file name for the output vector data (e.g., "statistic.shp") (Fig. 7.74).

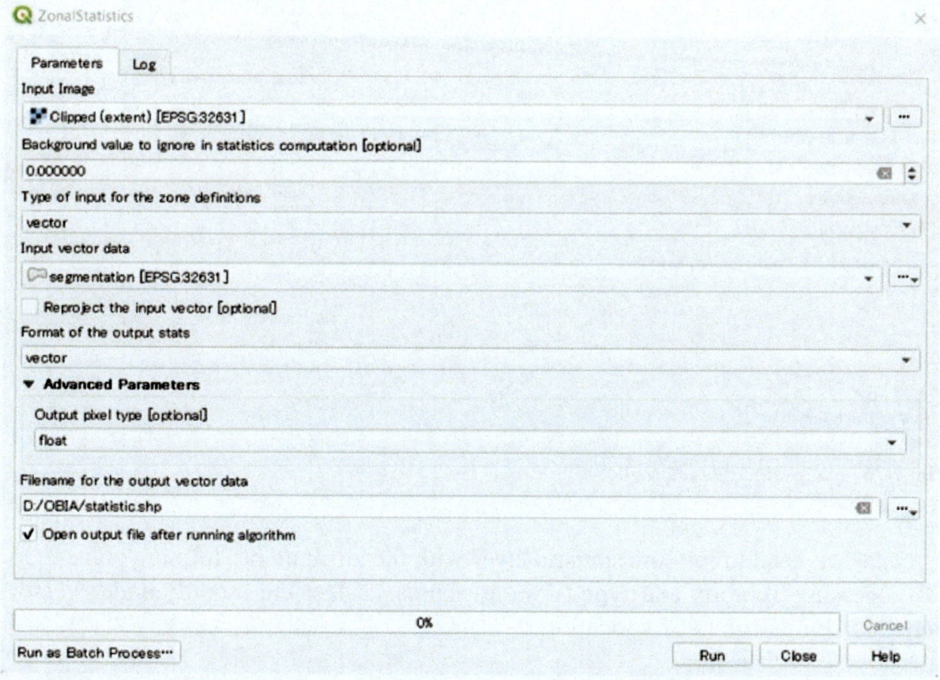

Fig. 7.74 Zonal statistic dialog box of OTB

In order to prevent encountering any error messages during the stage, it is necessary to rectify the geometry of the statistics.shp layer. To accomplish this, navigate to the Processing Toolbox and search for "Fix geometries" module, which can be found under Vector geometry. Select the module and execute it.

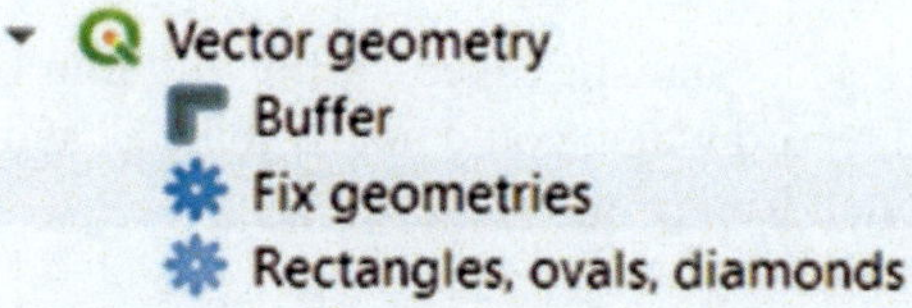

In the input layer, choose the layer "statistics," in the Fixed geometries type "statistic_fixed," and click Run button to initiate the process (Fig. 7.75).

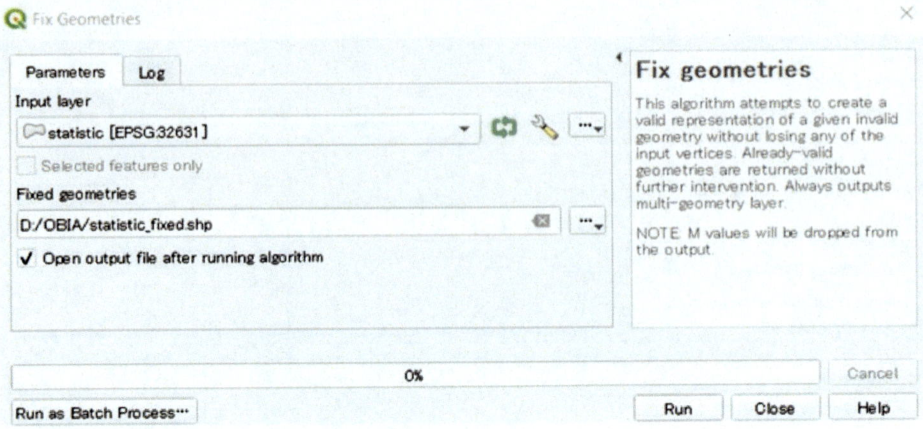

Fig. 7.75 Fix geometries dialog box

Next, we need to join the statistic layer with the attribute of "training" layer. Go to Processing Toolbox and type "Join attributes," select the module under Vector general > Join attributes by location."

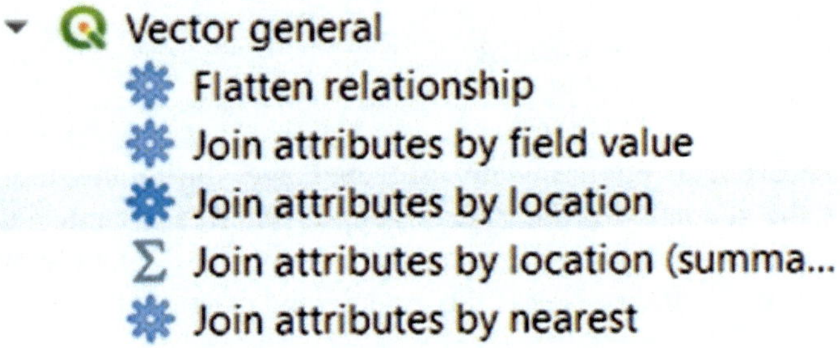

In the "Base Layer," select "statistic_fixed," while in "Join Layer" select "training." Check on "Discard record which could not be joined," and in the Joined layer, type "joined.shp" then click Run (Fig. 7.76).

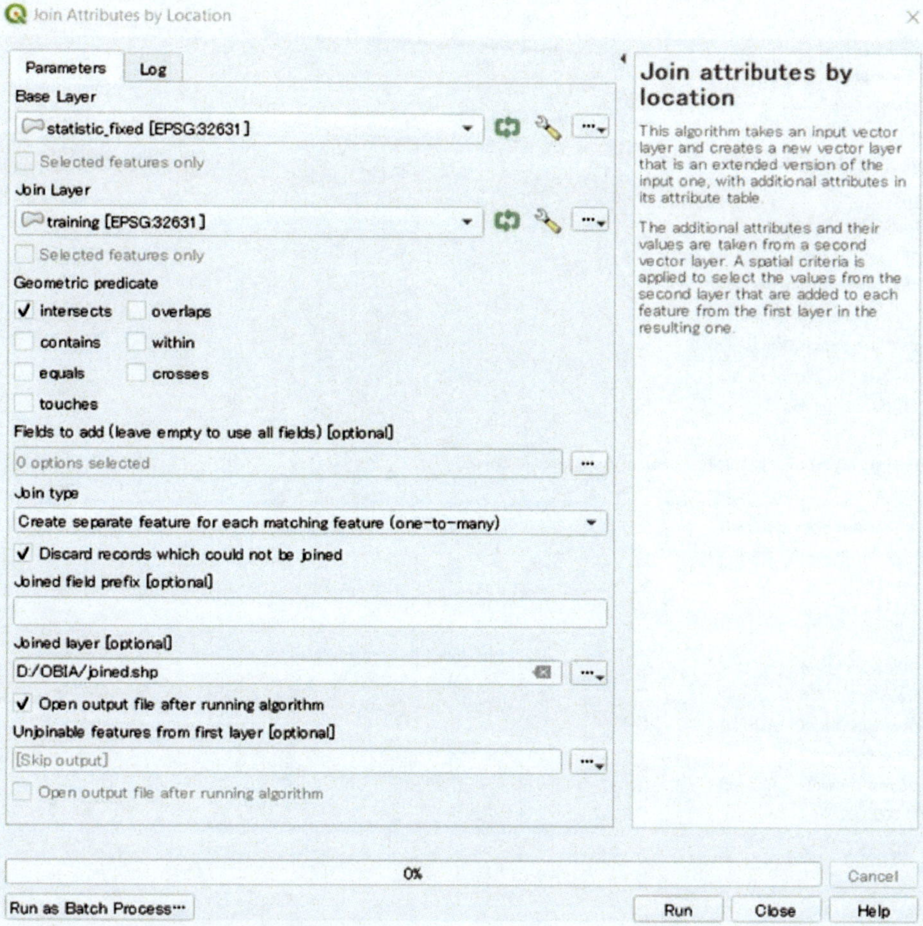

Fig. 7.76 Join attribute by location

Next step is training the vector classifier, go back to Processing Toolbox and select "TrainVectorClassifier" under OTB > Learning.

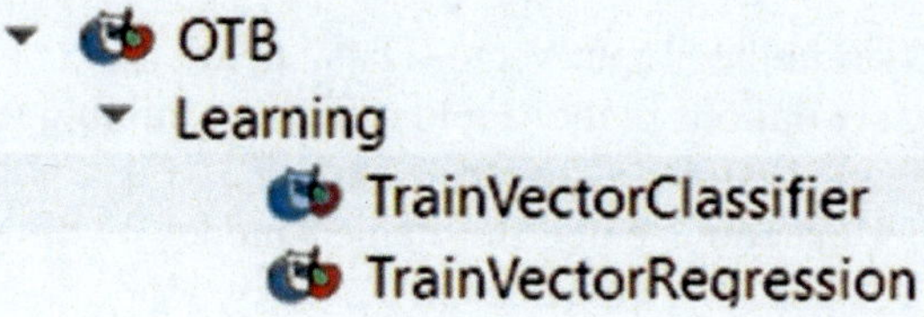

Fig. 7.77 Train vector classifier dialog box

To configure the input and output data for the model (Fig. 7.77), follow these steps:

1. Select the "joined" layer in the "Input Vector Data" field.
2. Choose the following six options in the "Field names for training features" field: **mean_0, stdev_0, mean_1, stdev_1, mean_2, stdev_2.** (Fig. 7.78).
3. In the "Output model" section, enter the name "model" with the extension ".model".

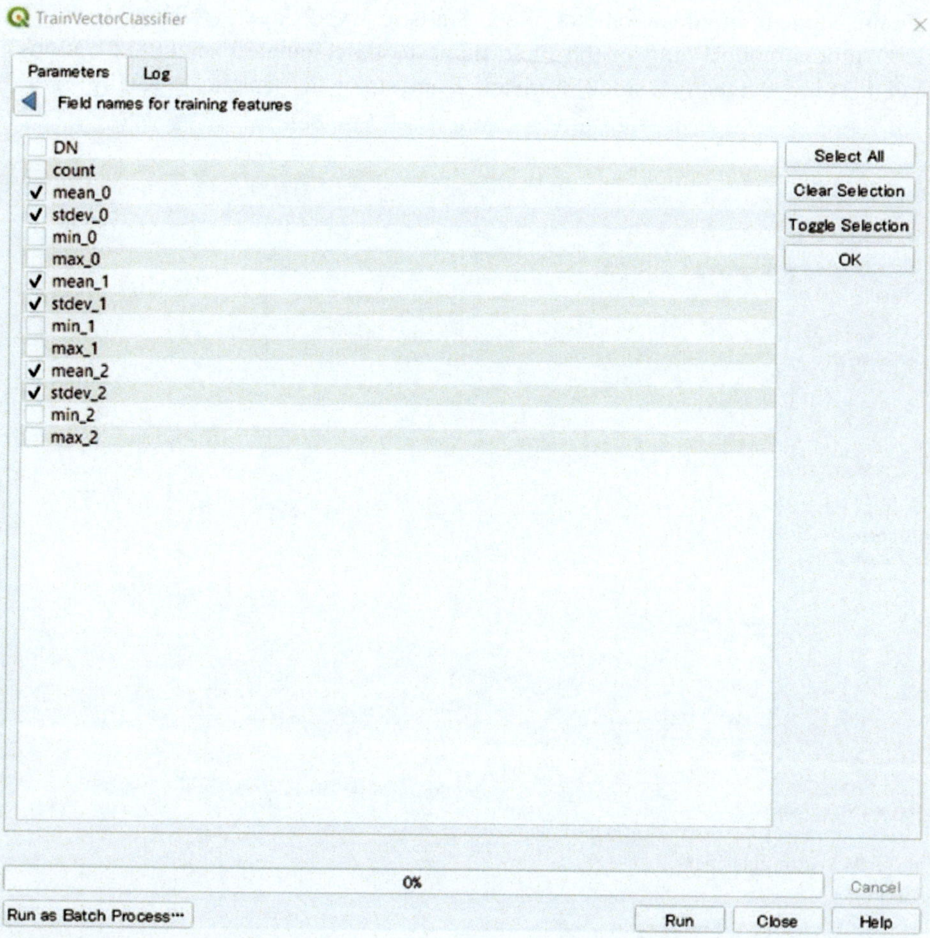

Fig. 7.78 Select the field names

The duration of the process will vary based on the specifications of your computer and the size of the satellite data. Typically, it should finish within a few minutes. Once the process is completed, navigate to the Processing Toolbox and search for "VectorClassifier." From the options presented, select the module located under OTB > VectorClassifier.

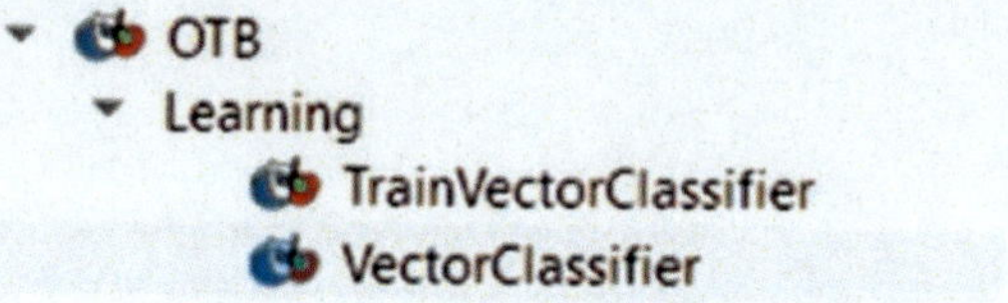

In the Name of input vector data select "statistic_fixed" layer, while in Model file select "model.model," and in the Field name to be calculated select six options: mean_0, stdev_0, mean_1, stdev_1, mean_2, and stdev_2, then click OK (Fig. 7.79).

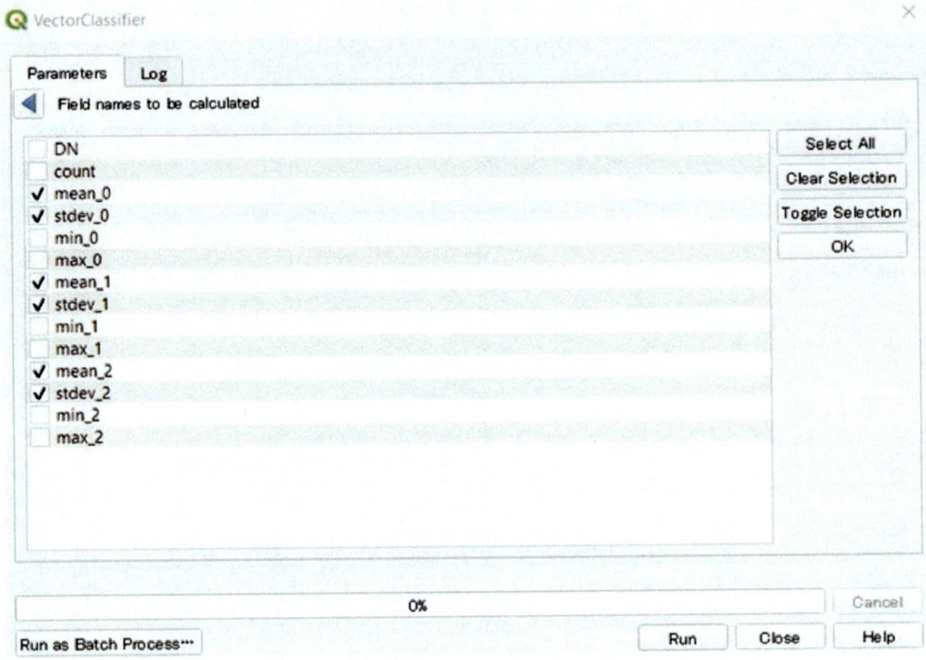

Fig. 7.79 Vector classifier

Ensure that the output vector data file is saved as "classified.shp" with the correct file extension (Fig. 7.80).

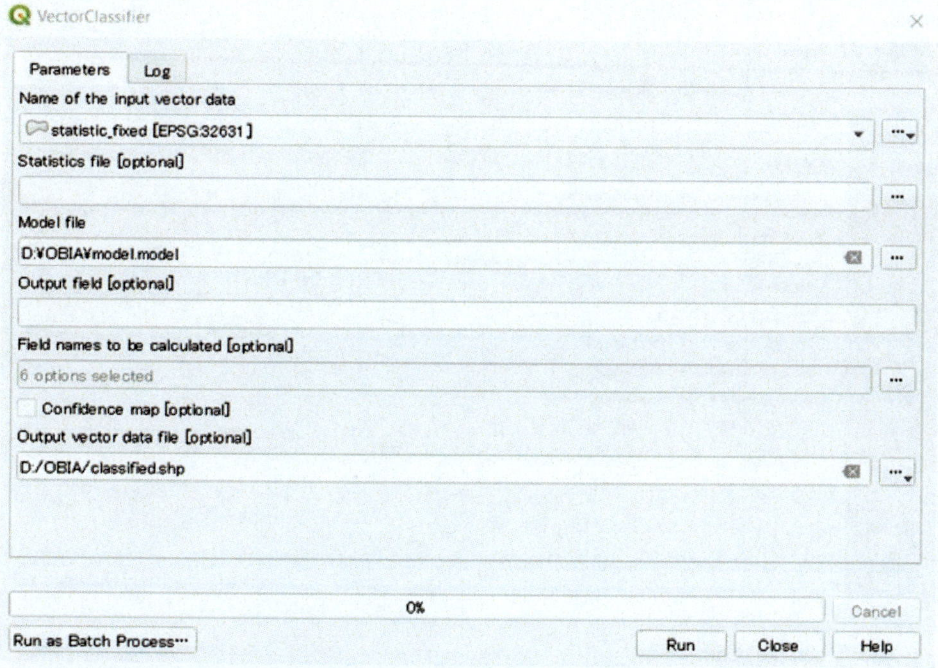

Fig. 7.80 Type the correct file extension

Failure to do so will result in an error notification as shown in Fig. 7.81.

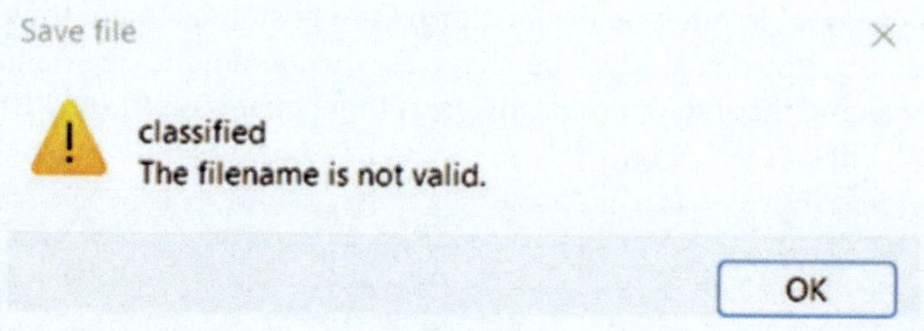

Fig. 7.81 Error notification

Then click Run to get the classification result.

After the processing is complete, the classification result will not be displayed automatically. To view the result, follow these steps: Go to Layer > Add Layer > Add Vector Layer (as shown in Fig. 7.82) and click on "Add."

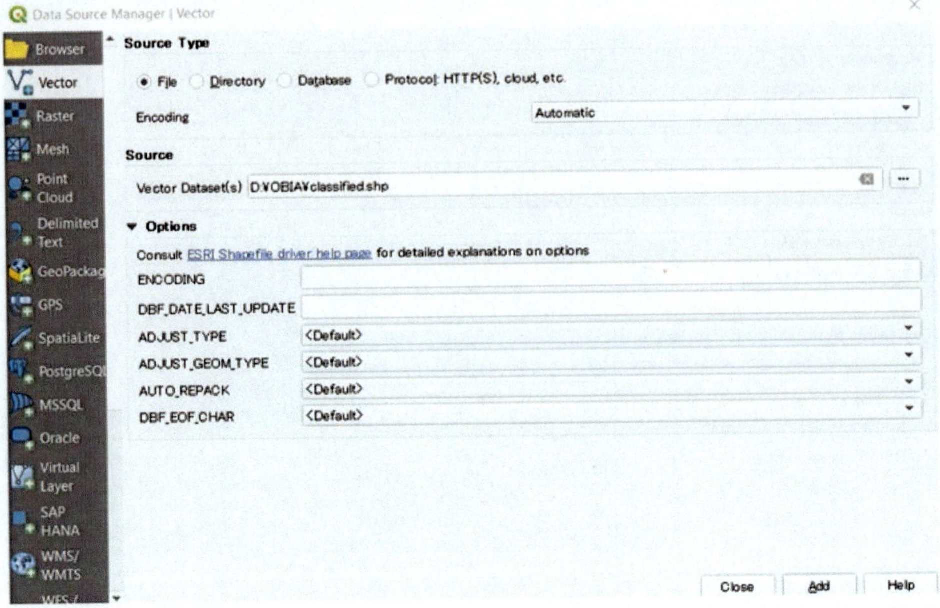

Fig. 7.82 Open the result

To view the results in categorized symbols, we need to adjust the symbology. To do this, double-click on the "classified" layer and select "Symbology." Change the display from "Single symbol" to "Categorized" and choose "predicted" in the "Value" field. Click "Classify" to generate categories. The colors may appear random, so you can double-click on each color box to customize them to your preference. Additionally, you can modify the "Legend" by double-clicking on the numeric labels and replacing them with descriptive text (e.g., changing "1" to "Buildings"). Once finished, click "OK" (Fig. 7.83).

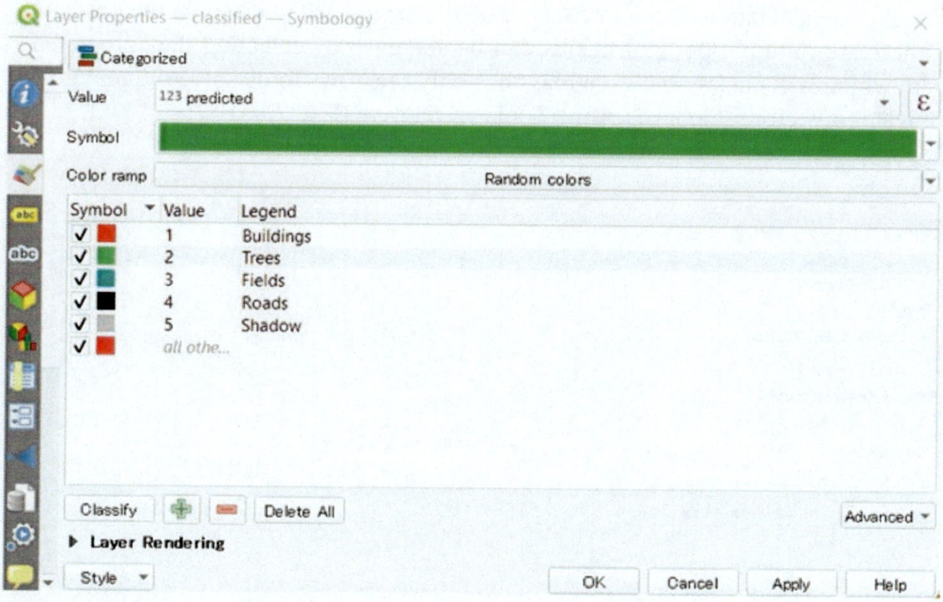

Fig. 7.83 Change the symbology of classification result

The final result of OBIA is as shown in Fig. 7.84.

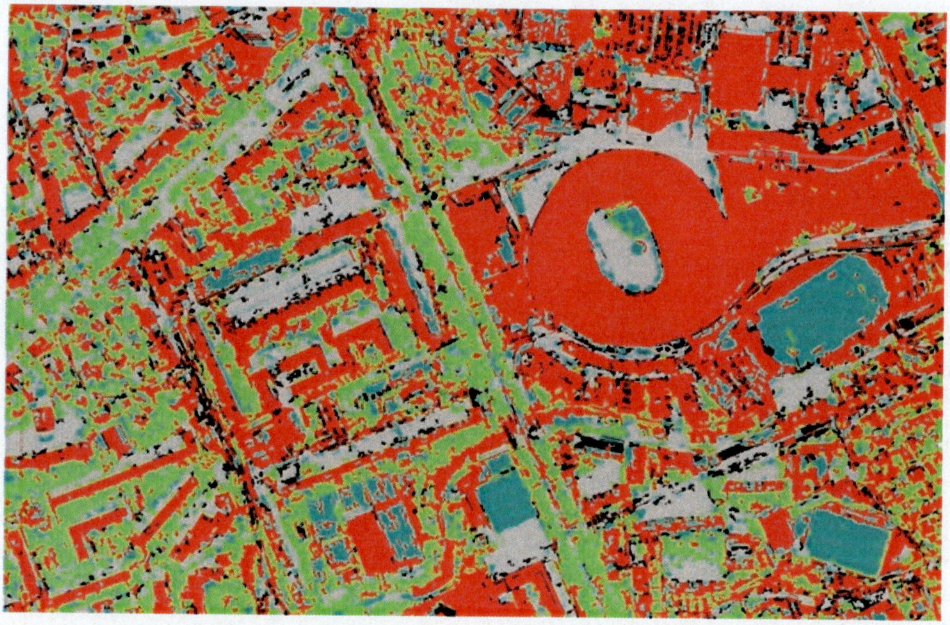

Fig. 7.84 Result of classification using OBIA method

Next, to calculate the area of each class, it is necessary to dissolve the polygons as there are multiple polygons. This can be achieved by selecting the "predicted" field, which will ensure that each class has only one row. To dissolve the polygons, navigate to Vector > Geoprocessing Tools > Dissolve (Fig. 7.85).

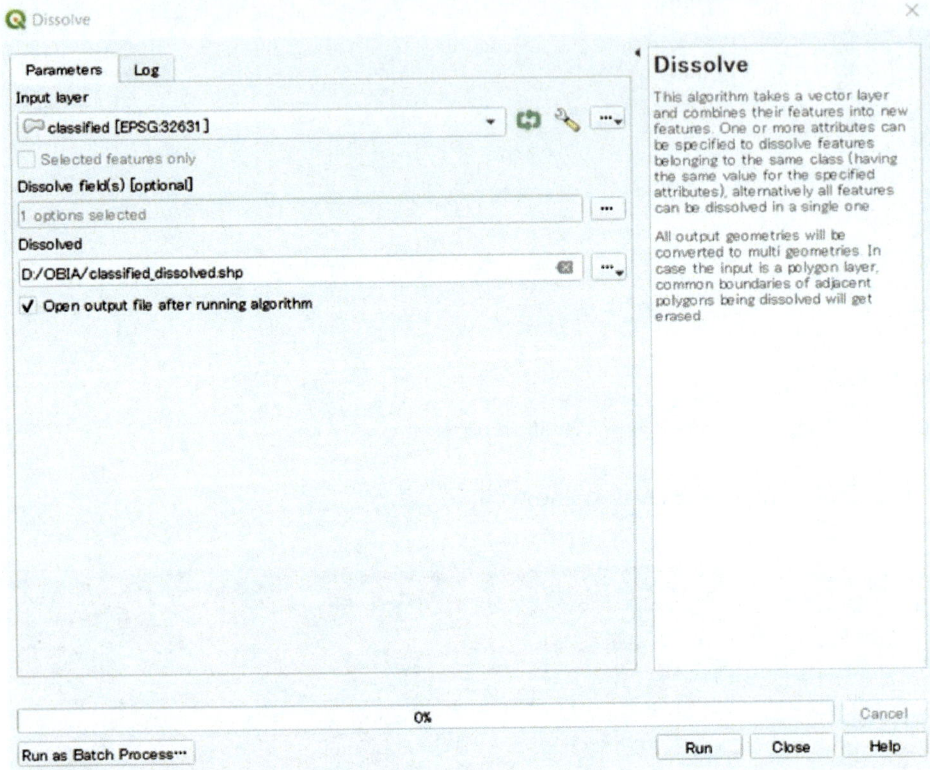

Fig. 7.85 The dissolve dialog box

In the Dissolve field(s) [optional] select "predicted," and select "Save to File" in the "Dissolved" field (Fig. 7.86).

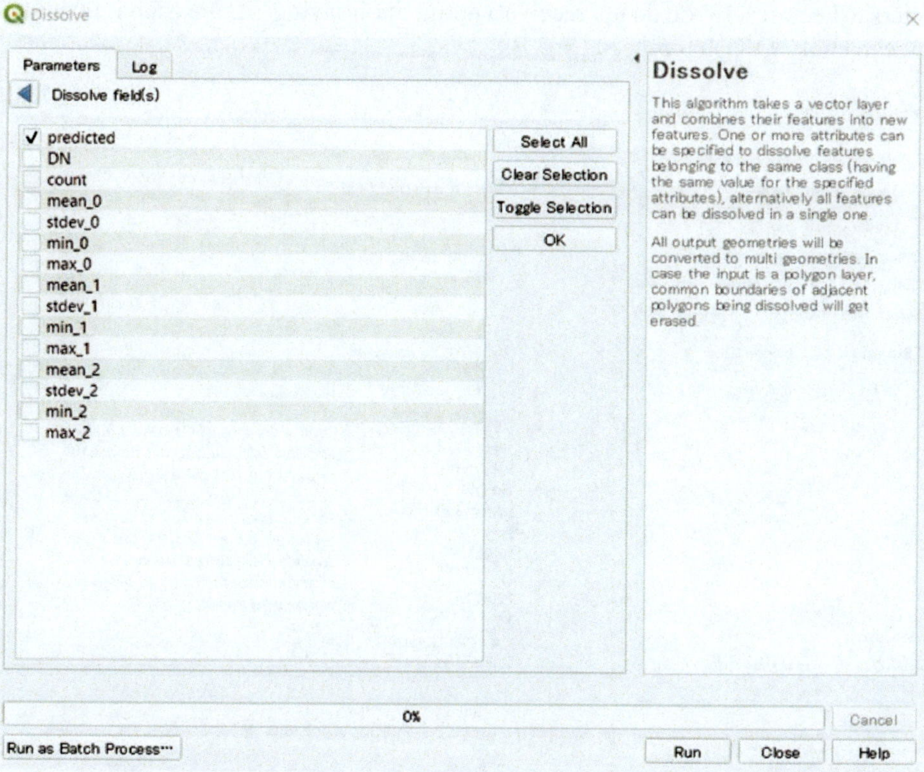

Fig. 7.86 Select the "predicted" field

After a couple of minutes, the result will be displayed on the main canvas of QGIS. To view the area in hectares instead of square meters, right-click on the "classified-dissolved" layer and select "Open Attribute Table." Then, click on the "Field Calculator" icon  and type "area(ha)" in the "Output field name" section. Next, choose "decimal number (real)" as the "Output field type" and type "$area*1e-4" in the "Expression" window. This will convert the unit from square meters to hectares. If you do not receive a notification saying "Expression is invalid," you can click "OK" to proceed (Fig. 7.87).

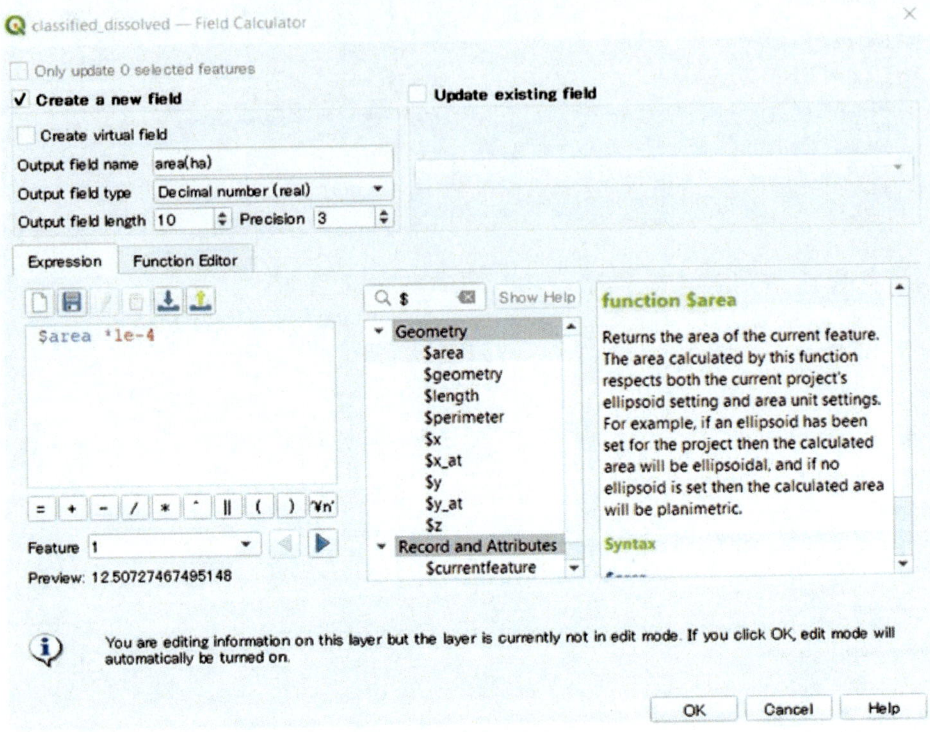

Fig. 7.87 Create a new field (ha)

Afterward, a new field titled "area(ha)" will be visible in the attribute table, as illustrated in Fig. 7.88.

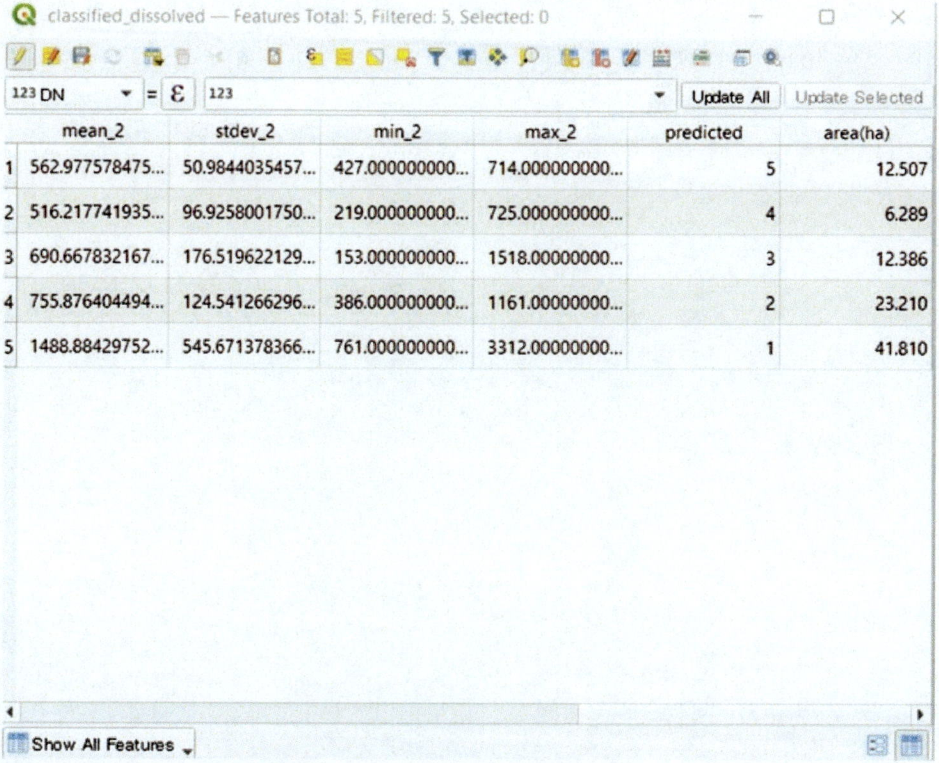

	mean_2	stdev_2	min_2	max_2	predicted	area(ha)
1	562.977578475...	50.9844035457...	427.000000000...	714.000000000...	5	12.507
2	516.217741935...	96.9258001750...	219.000000000...	725.000000000...	4	6.289
3	690.667832167...	176.519622129...	153.000000000...	1518.00000000...	3	12.386
4	755.876404494...	124.541266296...	386.000000000...	1161.00000000...	2	23.210
5	1488.88429752...	545.671378366...	761.000000000...	3312.00000000...	1	41.810

Fig. 7.88 Attribute table of classification result

Chapter 8
Integration of Remote Sensing Data with Other Spatial Datasets

Integration of remote sensing data with other spatial datasets using QGIS involves combining different types of geospatial data to create a more comprehensive and informative map or analysis. QGIS is a powerful open-source software that can handle different types of data and provides many tools for data integration and analysis. This chapter provides the steps to integrate remote sensing data with other spatial datasets using QGIS.

8.1 Download and Import the Remote Sensing and Other Spatial Datasets into QGIS

Remote sensing data can be in different formats such as raster or vector data and can be obtained from various sources such as satellites, drones, or LiDAR. Other spatial datasets can include vector data such as point, line, or polygon features, or raster data such as elevation, land use, or soil maps. Table 8.1 summarized the spatial data available for free, which can be used within QGIS.

F. Ramdani, *Exploring the Earth with QGIS*, Springer Remote Sensing/Photogrammetry, https://doi.org/10.1007/978-3-031-46042-5_8

Table 8.1 List of the spatial dataset available for free

Dataset name	Description and URL
Atmosphere/Ocean data	Data such as precipitation, soil moisture, vegetation, sea surface temperature are available at https://gportal.jaxa.jp/gpr/search?tab=0
Global land cover data	Copernicus global land service with spatial resolution of 100 m for year 2015 to 2019. Available at https://zenodo.org/record/3939038
Malaria	Global malaria datasets that can be downloaded based on the country level. Available at https://data.malariaatlas.org/maps
Natural earth	Basemap in vector type, available in multiple scale and can be downloaded at https://www.naturalearthdata.com/downloads/
Nighttime satellite data	Captured earth at night that can be search and download from https://eogdata.mines.edu/products/vnl/
PM2.5	Multitemporal data of PM2.5 from year 1998 to 2019. Available at https://sedac.ciesin.columbia.edu/data/set/sdei-global-annual-gwr-pm2-5-modis-misr-seawifs-aod-v4-gl-03/data-download
Population data	World population data that can be downloaded based on country name. Available at https://hub.worldpop.org/geodata/listing?id=32
Socioeconomic data and application center	Multiple socioeconomic data type and source. Available at https://sedac.ciesin.columbia.edu/

Note: This is not an exhaustive list and there may be other spatial datasets available for free

When integrating other spatial datasets with remote sensing data, there are several key points to consider, which include the following:

- Ensure that all the datasets have the same projection and are properly aligned. This can be done by checking the properties of each layer and reprojecting them if necessary.
- Combine the datasets using various spatial analysis tools provided in QGIS. Some common methods of integration include overlaying layers, merging datasets, or joining attribute tables.
- Conduct analysis using the integrated datasets to gain insights or to create new data products. This can include identifying areas of land cover change, calculating vegetation indices, or creating terrain models.
- Visualize the results using QGIS's mapping and visualization tools. This can include creating thematic maps, adding labels or annotations, or exporting the results to various formats for further use.

Integration of remote sensing data with other spatial datasets using QGIS can provide a more complete picture of Earth's surface and help to identify patterns and trends that may not be visible with individual datasets alone.

8.2 Integrating with LiDAR

LiDAR stands for Light Detection and Ranging, which is a remote sensing technology that uses laser pulses to measure distances to Earth's surface. It can create precise 3D models of the terrain, buildings, and vegetation by measuring the time it takes for the laser beam to bounce back from the surface. LiDAR can capture millions of data points per second, providing highly detailed information about the surface.

The importance of LiDAR lies in its ability to accurately capture and represent topographical features and changes in elevation. This information is crucial for a variety of applications such as urban planning, forestry management, flood modeling, and infrastructure design. LiDAR can also be used for archaeological surveys, coastal zone mapping, and even self-driving cars.

QGIS can be used to analyze and process LiDAR data. Here are the basic steps to use LiDAR in QGIS:

- Import LiDAR data into QGIS: LiDAR data is usually provided in the form of LAS files, which can be loaded into QGIS using the WhiteboxTool plugin.
- Visualize the LiDAR data: QGIS provides several tools to visualize the LiDAR data.
- Extract terrain information: QGIS provides tools to extract terrain information from LiDAR data, such as the Terrain Analysis plugin, which can generate elevation contours, slope maps, and aspect maps.
- Perform analysis: Once the terrain information has been extracted, QGIS can be used to perform various analysis such as flood modeling, viewshed analysis, and vegetation analysis.
- Export results: QGIS provides various options to export the results of the analysis, such as geoTIFFs.

LiDAR is an important technology that provides highly detailed information about Earth's surface, and it can be used in various applications. QGIS is a powerful tool that can be used to analyze and process LiDAR data, and it provides several tools to visualize, extract, and analyze terrain information.

To integrating with LiDAR, you need to download the WhiteboxTool (available at: https://www.whiteboxgeo.com/download-direct/) and extract it, then install the plugin from QGIS (Fig. 8.1).

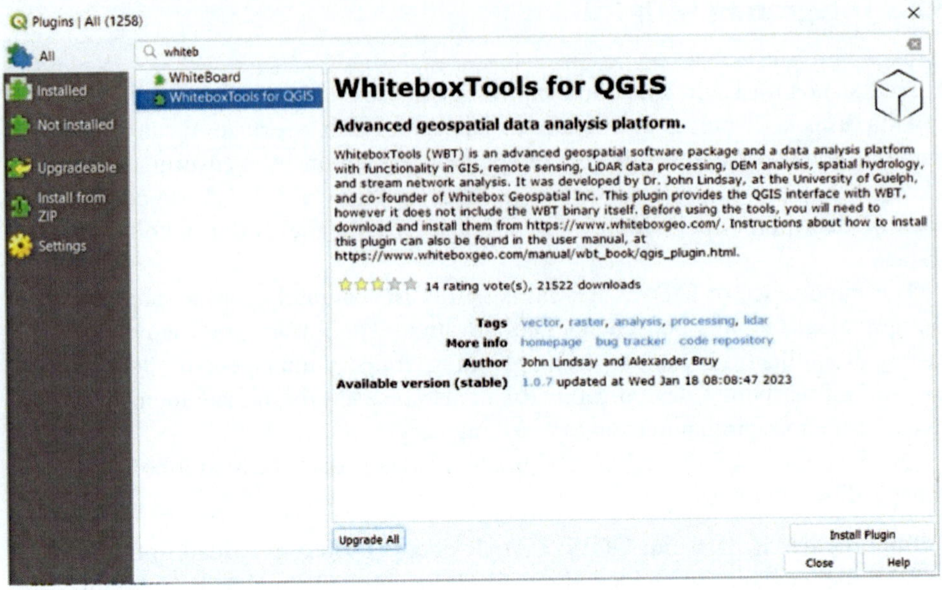

Fig. 8.1 Install WhiteboxTools for QGIS

Next, you need to change the setting and locate the path of the executable WBT from Processing Toolbox of QGIS (Fig. 8.2). Click on the icon .

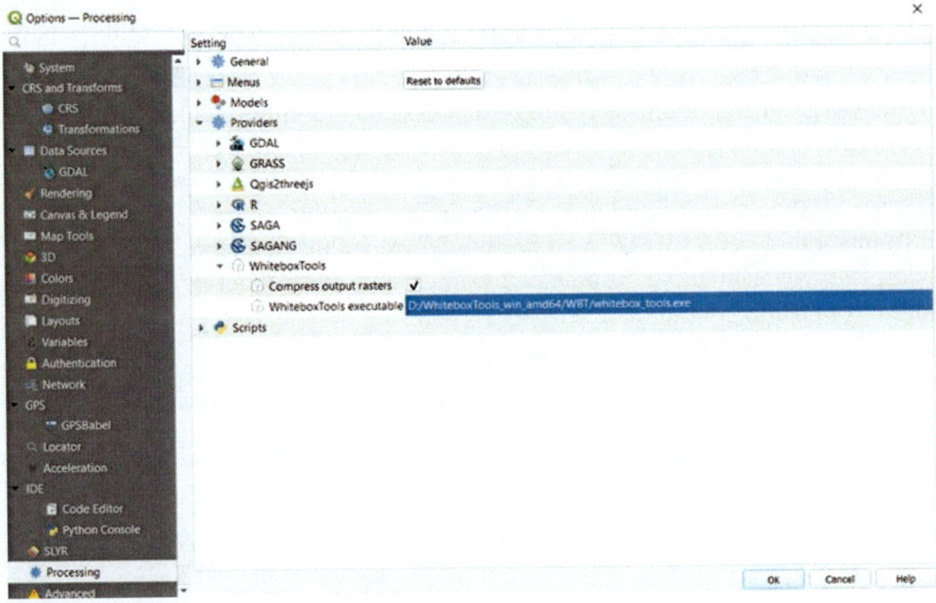

Fig. 8.2 Setting the path of WBT

For this practice you can find and download LiDAR data from Scottish Remote Sensing Portal (https://remotesensingdata.gov.scot/data#/map) (Fig. 8.3).

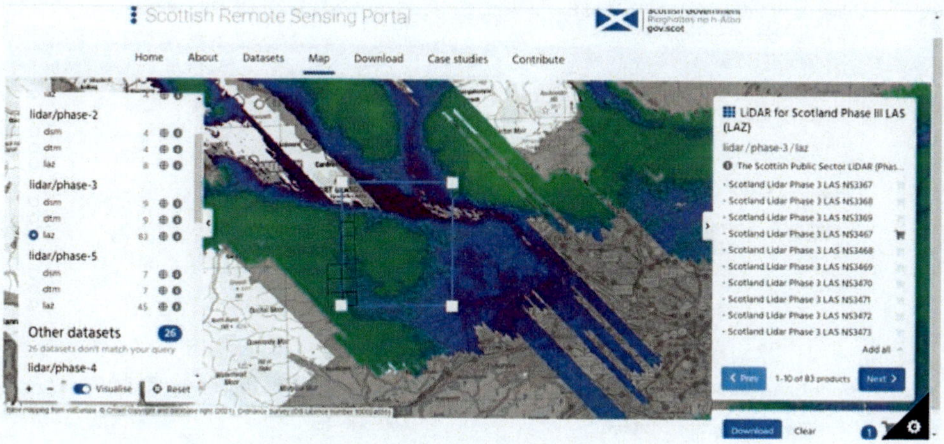

Fig. 8.3 The Scottish Remote Sensing Portal

Look for LiDAR/phase-3 then click on available data, click on add symbol and Download (Fig. 8.4).

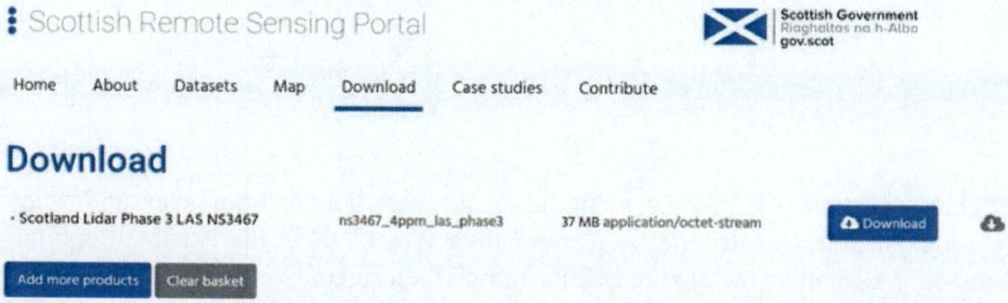

Fig. 8.4 Download LiDAR phase 3

Extract the downloaded data and open it using QGIS, go to Layer > Add Layer > Add Point Cloud Layer. Then click on icon next to point cloud layer's name, and assign the CRS to EPSG 27700 (Fig. 8.5).

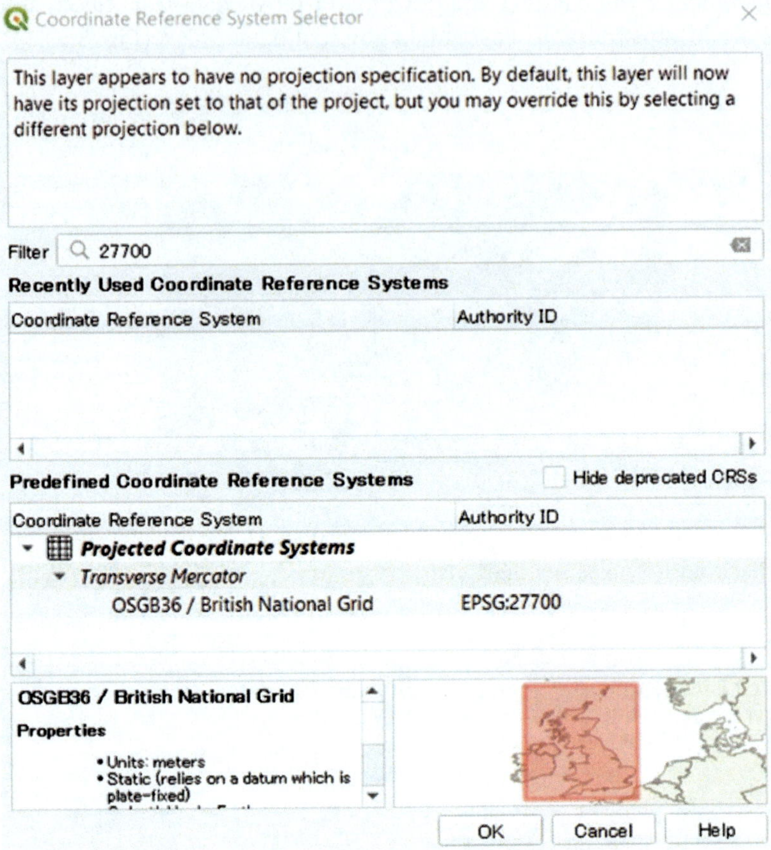

Fig. 8.5 Assign the CRS

Then you can assign new symbology, double-click on the layer and select Symbology. Select “Attribute by Ramp,” then select “Z” in the Attribute column. Click on the small arrow next to “Color ramp” then select Create new color ramp > Catalog: cpt-city. Then scroll down, click on Topography > elevation, then click OK-Apply (Fig. 8.6).

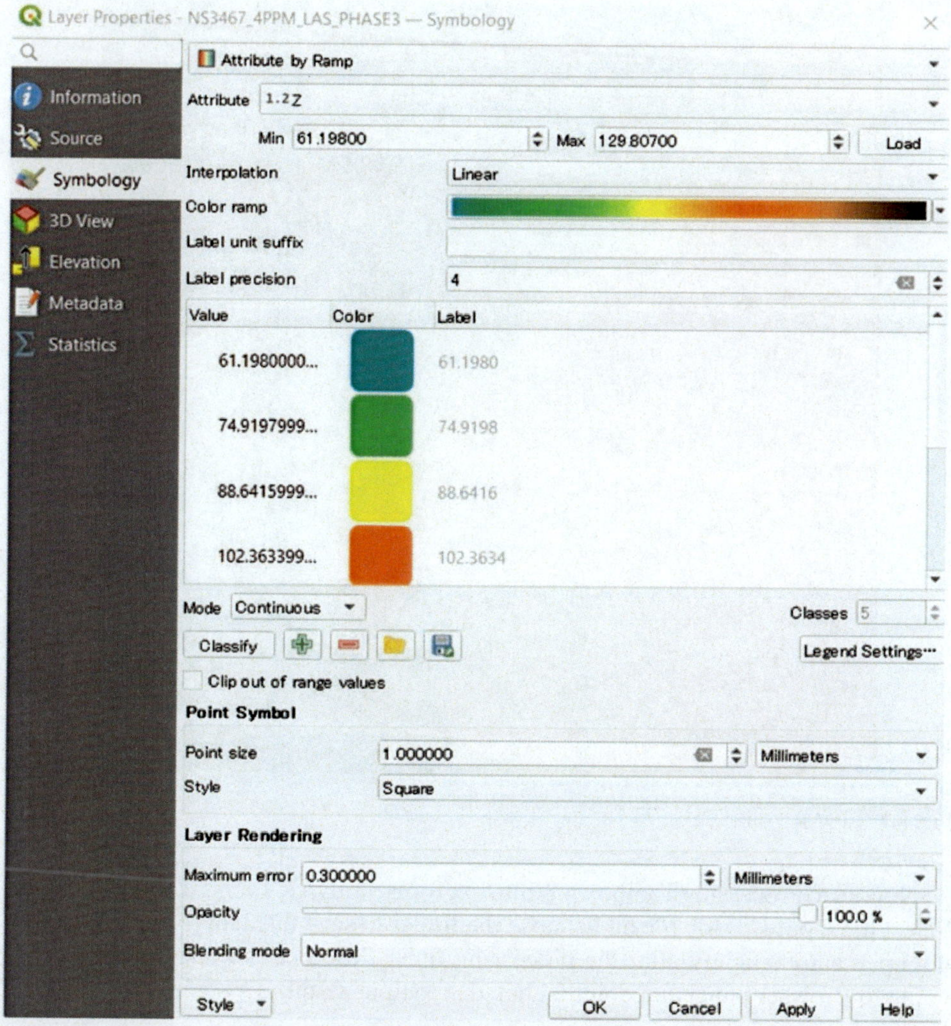

Fig. 8.6 Assign new symbology of LiDAR data

The result will be visualized as Fig. 8.7.

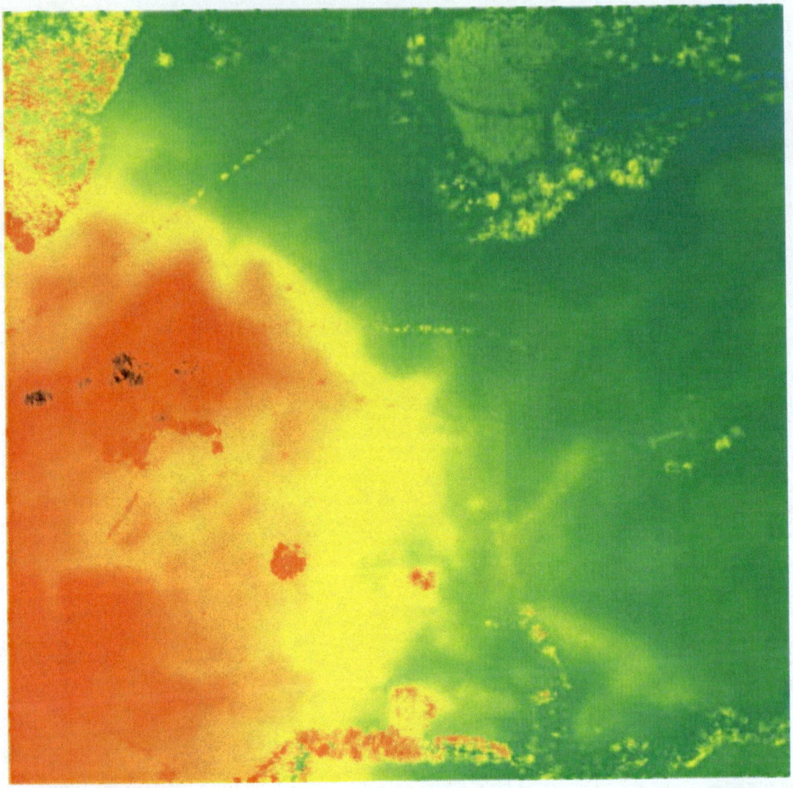

Fig. 8.7 LiDAR visualization

Now go to Processing Toolbox > WhiteboxTools > LiDAR Tools > LidarHillshade. Select the input LiDAR file and choose the folder to save the data (Fig. 8.8). It will not open automatically after the processing finish. You need to open manually by going to Browser panel and double-click data output result.

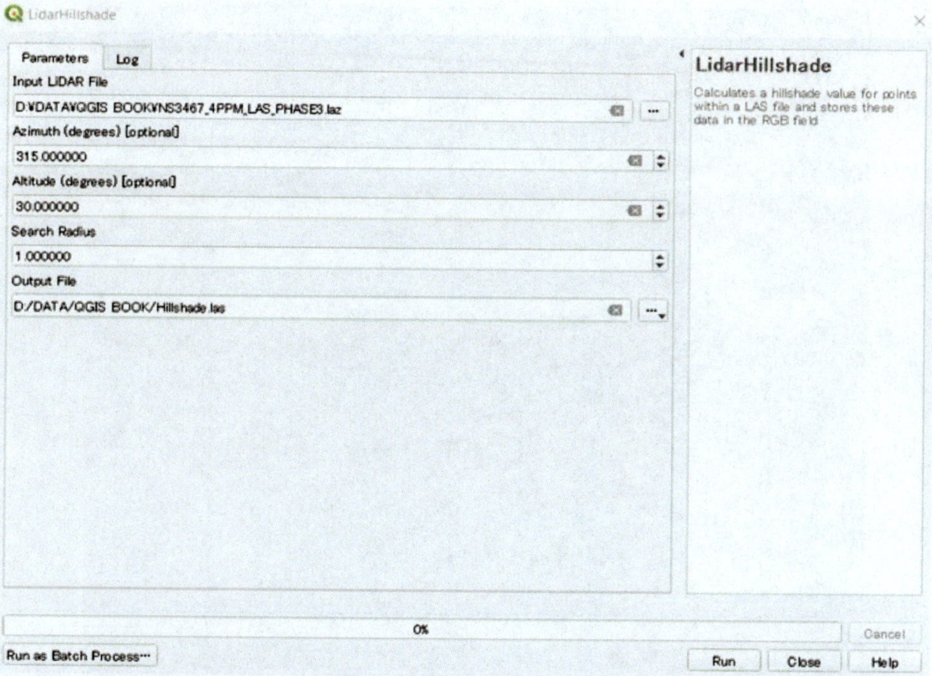

Fig. 8.8 LiDAR hillshade dialog box

If it shows nothing after you open the file, click on icon and assign the CRS to EPSG 27700. The result will be shown as Fig. 8.9.

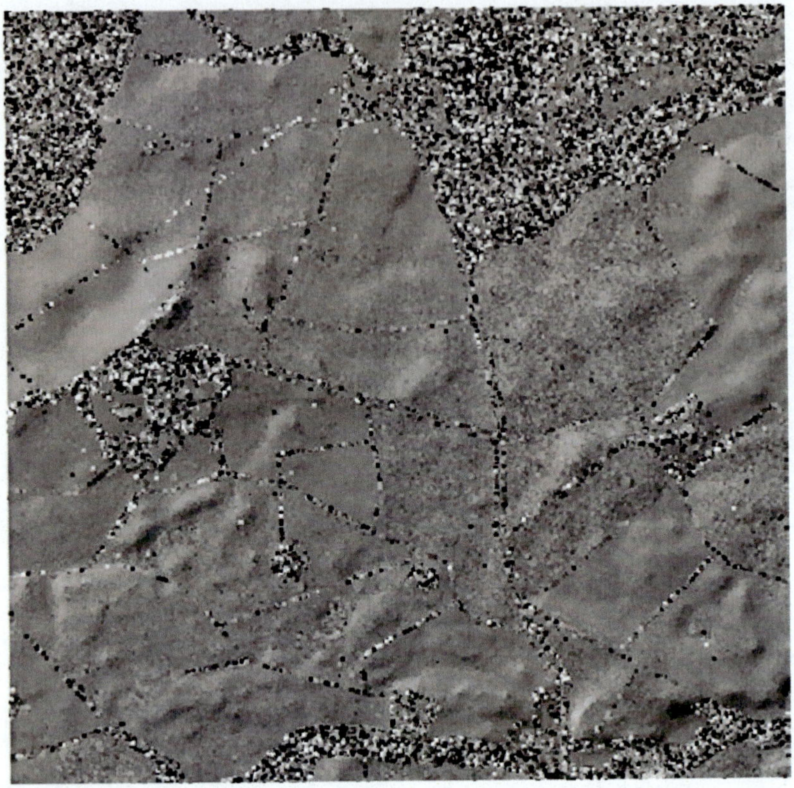

Fig. 8.9 Visualization of LiDAR hillshade

Next is to produce Digital Surface Model (DSM), go to Processing Toolbox > WhiteboxTools > LiDAR Tools > LidarDigitalSurfaceModel (Fig. 8.10).

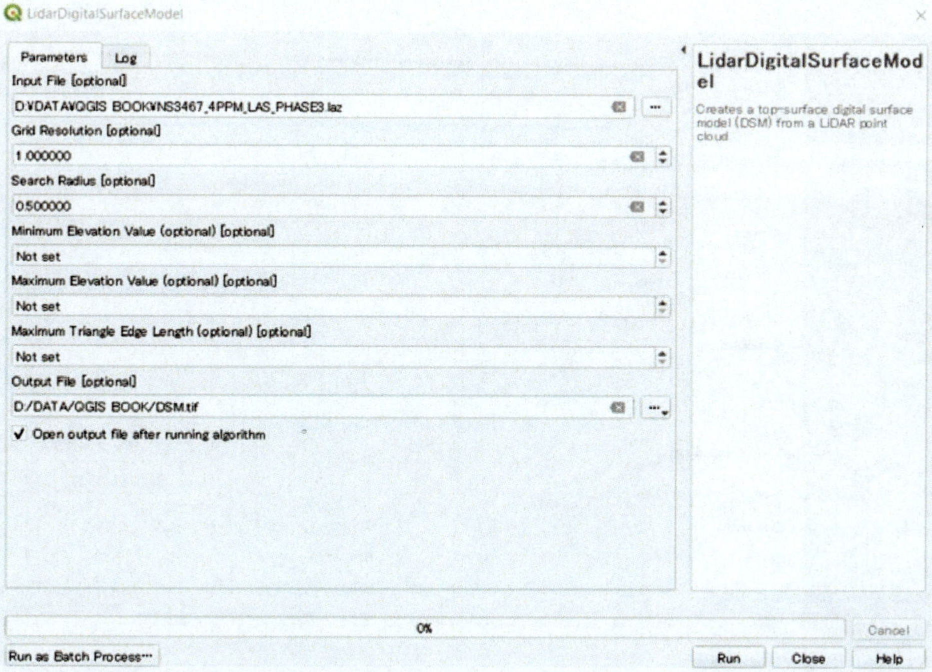

Fig. 8.10 LiDAR DSM dialog box

Choose the input file and select the folder to save the output file, then click Run. Right-click the output DSM layer then CRS to EPSG 27700. Then double-click, select Symbology, change Render Type to "Hillshade" and Blending mode to "Multiply" (Fig. 8.11).

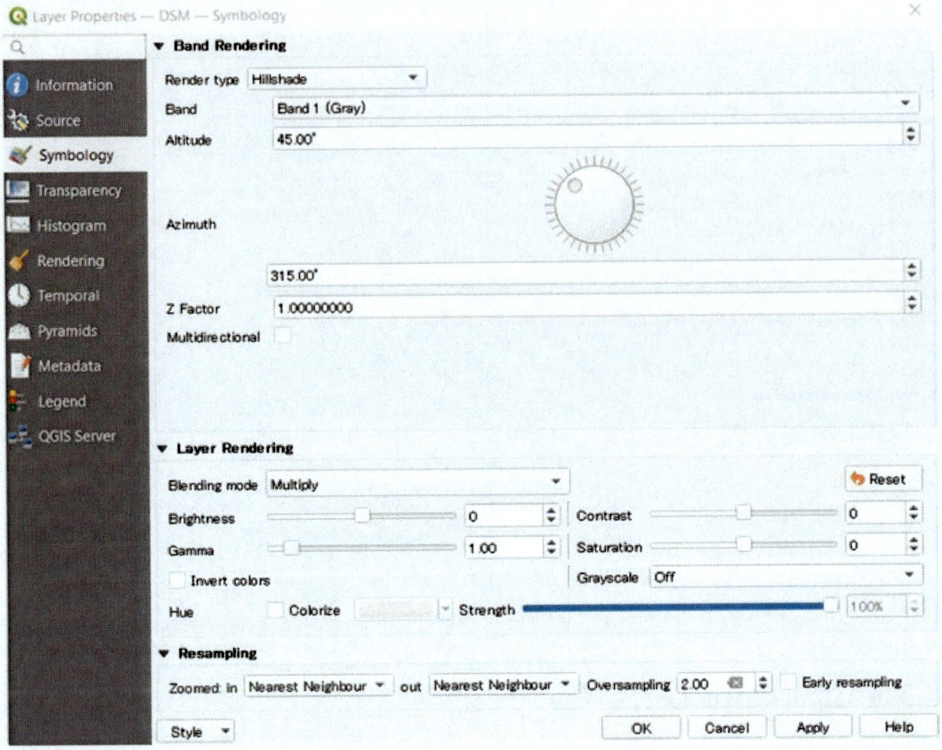

Fig. 8.11 DSM symbology dialog box

The final result will be as Fig. 8.12.

Fig. 8.12 Visualization of LiDAR DSM

We can also create 3D view, double-click on layer "NS3467_4PPM_LAS_PHASE3" then select 3D View, select "Attribute by Ramp," then select "Z" in the Attribute column. Click on the small arrow next to "Color ramp" then select Create new color ramp > Catalog: cpt-city. Then scroll down, click on Topography > elevation, then click OK-Apply (Fig. 8.13).

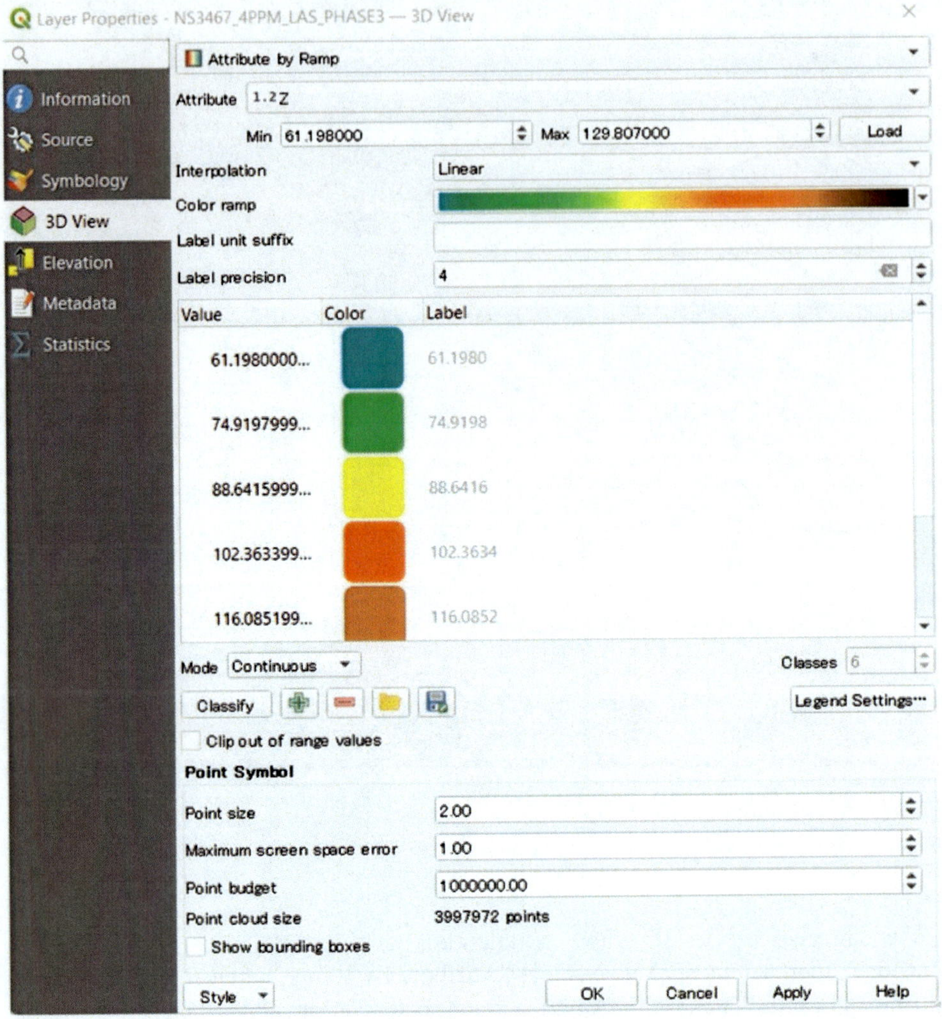

Fig. 8.13 3D view dialog box

Go to View - New 3D Map View. Change the projection if the notification is appearing such as Fig. 8.14.

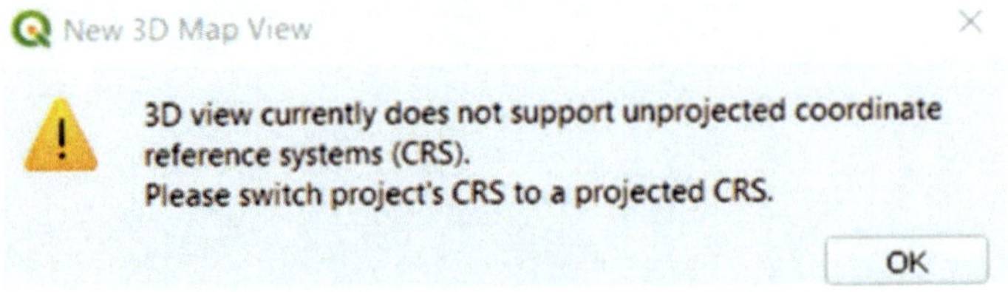

Fig. 8.14 Warning notification of unprojected CRS

Click on the icon of the EPSG:4326 located on the bottom-right of QGIS window, then change to EPSG 27700.

On the 3D Map window, click on the configuration icon, then change Type to "DEM" (Raster Layer), Elevation to "DSM," and check on Terrain Shading, then click OK (Fig. 8.15).

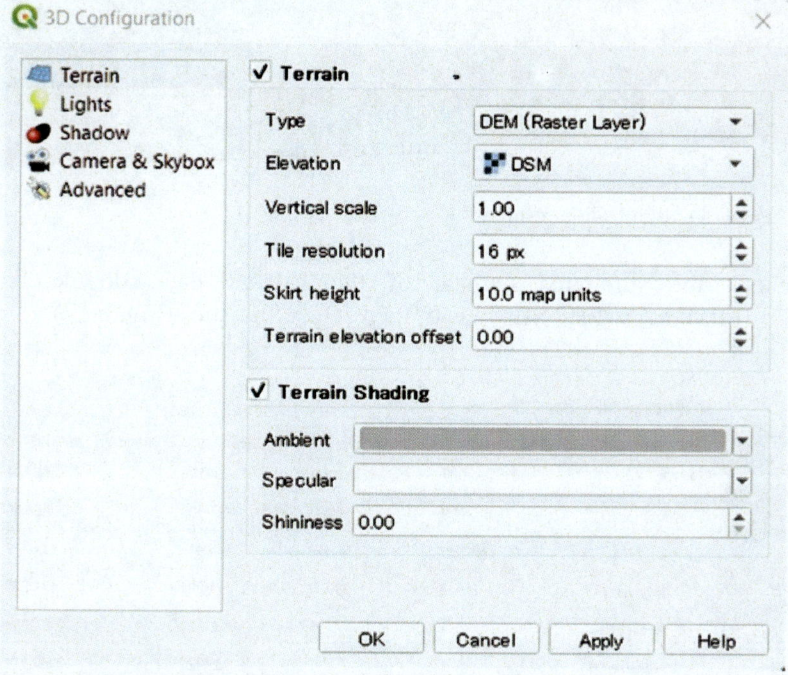

Fig. 8.15 3D configuration dialog box

To navigate click on the icon of "Toggle on Screen Navigation" then make some adjustment. After some adjustment, the final result will be as Fig. 8.16.

Fig. 8.16 Final 3D visualization result

To check the summary statistic of the LiDAR data, double-click on "NS3467_4PPM_LAS_PHASE3" layer, then select Statistics (Fig. 8.17).

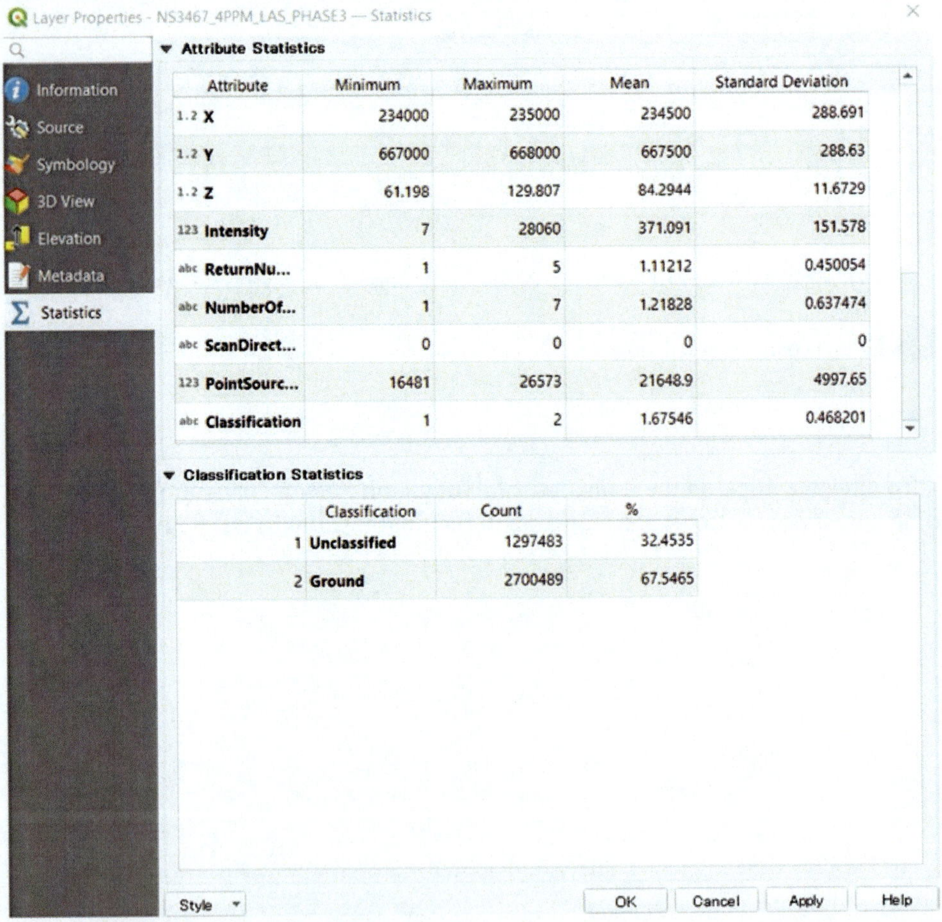

Fig. 8.17 Attribute statistics of LiDAR layer

Chapter 9
Integration of QGIS with Google Earth Engine (GEE)

Google Earth Engine (GEE) is a cloud-based platform that provides access to petabytes of remote sensing data and allows users to perform advanced geospatial analysis. Integrating QGIS with GEE can provide users with additional tools and resources for working with geospatial data. There are a few different ways to integrate QGIS with GEE, including using plugins and accessing GEE data directly through the QGIS interface.

One popular plugin for integrating QGIS with GEE is the "GEE Plugin." This plugin allows users to access GEE datasets, perform geospatial analysis using GEE algorithms, and import the results back into QGIS. The plugin also provides a visual interface for exploring GEE datasets and selecting the specific data layers that you want to work with.

Another way to integrate QGIS with GEE is by using the GEE Python API. This API allows users to write Python scripts that can interact with GEE datasets and perform geospatial analysis. QGIS provides a Python console where users can write and execute Python scripts, making it easy to work with GEE data within the QGIS environment.

9.1 Advantages and Disadvantages

When working with GEE data in QGIS, there are a few key benefits to keep in mind. One of the biggest advantages is the ability to combine GEE data with other geospatial datasets that you may have stored locally. This can help you to create more comprehensive and detailed analyses that take into account a wider range of variables.

Another advantage of using GEE data within QGIS is the ability to perform more complex geospatial analyses than would be possible with local data alone. GEE provides access to a wide range of remote sensing data, which can be used to

F. Ramdani, *Exploring the Earth with QGIS*, Springer Remote Sensing/Photogrammetry, https://doi.org/10.1007/978-3-031-46042-5_9

analyze land cover, vegetation health, and other environmental variables. By combining this data with other geospatial datasets, you can create detailed maps and analyses that provide valuable insights into the environment.

Integrating QGIS with GEE can also help to streamline your workflow and make it easier to manage and analyze large geospatial datasets. With the ability to access GEE data directly through the QGIS interface, you can avoid the need to download and store large amounts of data locally. This can save time and storage space, while also reducing the risk of data loss or corruption.

However, there are a few key considerations to keep in mind when working with GEE data within QGIS. One of the most important is ensuring that you have a reliable and fast Internet connection. GEE data is stored in the cloud, which means that you need a stable and speedy Internet connection. Furthermore, the downloaded data is only for visualization purposes, you will not be able to do analysis based on the data.

9.2 Install and Authentication Process

This chapter will introduce you how to using python console within QGIS, access, and then visualize the data using API of GEE. Before doing that you need to register to GEE first using the following URL: https://earthengine.google.com/new_signup

After finishing the registration, you can install the GEE Plugin (Fig. 9.1).

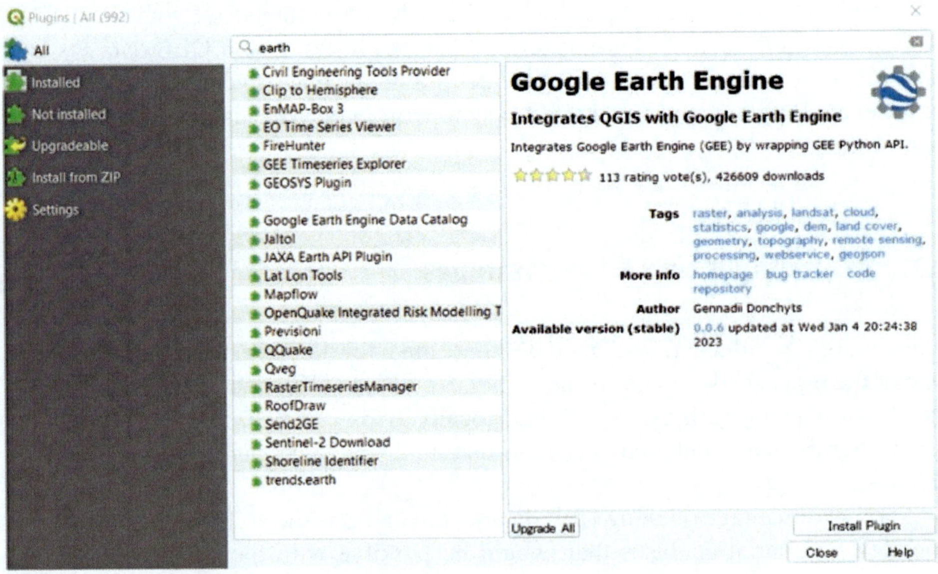

Fig. 9.1 Install GEE plugin

The first time you use it, it will open a new window that informs you as follows:

This plugin uses Google Earth Engine API and it looks like it is not yet authenticated on this machine. You need to have a Google account registered in Google Earth Engine to continue
Click OK to open a web browser and start the authentication process or click Cancel to stop the authentication process.

Just click OK then follow the steps to login into your GEE Account, then click "Allow" (Fig. 9.2).

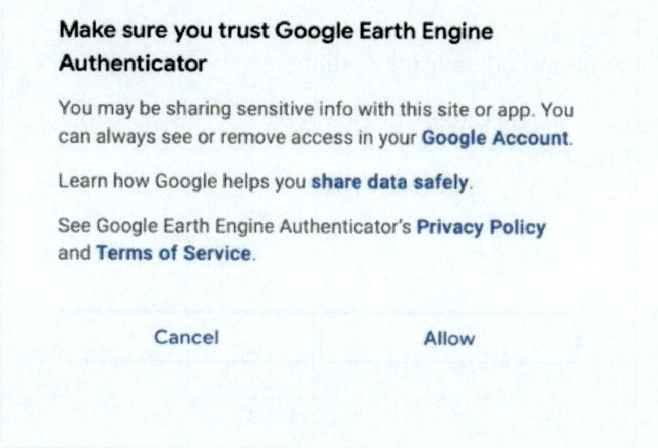

Fig. 9.2 GEE authenticator

9.3 Visualizing Satellite Images

9.3.1 Sentinel-2

Sentinel-2 data is important for a variety of applications in the fields of Earth observation, remote sensing, and environmental monitoring. Here are some key reasons why:

- Land cover and land use mapping: Sentinel-2 data provides high-resolution multispectral imagery, which is ideal for mapping and monitoring land cover and land use changes over time. This is useful for various applications, such as monitoring urbanization, deforestation, and agricultural practices.
- Natural resource management: The data provided by Sentinel-2 is useful for managing natural resources, such as water and forests. The data can be used to monitor water quality, detect changes in forest cover and health, and monitor vegetation health and productivity.
- Disaster management: Sentinel-2 data can be used to monitor and respond to natural disasters, such as floods, wildfires, and earthquakes. The data can help in

mapping the extent of the disaster, identifying areas that require immediate attention, and aiding in disaster response and recovery efforts.
- Climate change monitoring: Sentinel-2 data is useful for monitoring the effects of climate change on the environment. The data can be used to track changes in the distribution and abundance of plant species, monitor changes in snow and ice cover, and monitor changes in Earth's surface vegetation.
- Urban planning: Sentinel-2 data can be used for urban planning and management. The data can help in mapping urban growth and development and monitoring land use patterns.

To visualize Sentinel-2 data using python, open your QGIS click on icon to open the python console, then click on icon to show the editor. Copy and paste the following code:

```
import ee
from ee_plugin import Map

#Defining study area
boundary = ee.FeatureCollection("USDOS/LSIB_SIMPLE/2017");
country = boundary.filterMetadata('country_na','equals','Nigeria');

#Defining the datasets and clip to study area
data = ee.ImageCollection("COPERNICUS/S2_HARMONIZED")\
        .filterDate("2022-01-01", "2022-03-31")\
        .filterBounds(country)
s2composite = data.median().clip(country)

#Visualization parameters
trueColour = {'bands': ["B11", "B8", "B2"], 'min': 0, 'max': 6000,
'gamma': 1.5}

#Map and display
Map.addLayer(s2composite, trueColour, "Nigeria-Sentinel, 2022")
Map.centerObject(country, 7)
```

Then click run icon it will visualize the satellite image of Nigeria using Sentinel-2 image (Fig. 9.3).

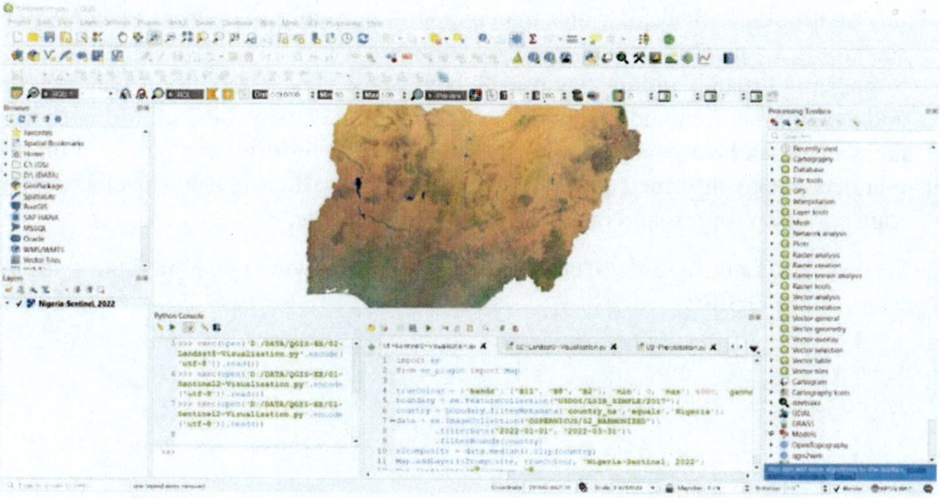

Fig. 9.3 Sentinel-2 image of Nigeria displayed within QGIS using GEE plugin

Click on the save icon  to save your code.

9.3.2 *Landsat-8*

Landsat-8 is a satellite mission jointly operated by NASA and the US Geological Survey (USGS) that provides valuable data on Earth's land surface. Here are some important reasons why Landsat-8 data is significant:

- Monitoring land use and land cover changes: Landsat-8 captures images of Earth's surface at a spatial resolution of 30 meters, which allows researchers and analysts to monitor changes in land use and land cover over time. This information is vital for understanding how human activities, such as deforestation, urbanization, and agriculture, are impacting Earth's ecosystems.
- Tracking natural disasters: Landsat-8's ability to capture images of Earth's surface on a regular basis makes it a valuable tool for tracking natural disasters such as floods, wildfires, and landslides. These images can help emergency responders and relief organizations quickly assess the extent of damage and plan response efforts.
- Supporting agricultural planning: Landsat-8 data can be used to monitor crop growth, crop health, and crop yields, which is essential for agricultural planning and management. This information can help farmers decide when to plant, irrigate, fertilize, and harvest their crops.
- Assessing water resources: Landsat-8 data can be used to monitor water resources, including changes in water levels, water quality, and water usage. This

information is critical for water management, particularly in areas where water scarcity is a concern.

- Supporting urban planning: Landsat-8 data can be used to monitor urban growth and development, including changes in population density, land use, identifying areas of urban heat islands, and infrastructure. This information can help urban planners make informed decisions about where to locate new developments, transportation routes, and other infrastructure projects.

To visualize Landsat-8 data, copy and paste the following code into your python console within QGIS:

```
import ee
from ee_plugin import Map

#Defining the study area
boundary = ee.FeatureCollection("USDOS/LSIB_SIMPLE/2017");
country = boundary.filterMetadata('country_na','equals','Nigeria');

#Defining the datasets and clip to study area
data = ee.ImageCollection("LANDSAT/LC08/C02/T1_TOA")\
        .filterDate("2022-01-01", "2022-03-31")\
        .filterBounds(country)
l8composite = data.median().clip(country)

#Visualization parameters
trueColour = {'bands': ["B4", "B3", "B2"], 'min': 0.02, 'max':
0.35, 'gamma': 1.5}

#Map and display
Map.addLayer(l8composite, trueColour, "Nigeria-L8, 2022")
Map.centerObject(country, 7)
```

Then click run icon ▶ it will visualize the satellite image. It will visualize the Landsat-8 of Nigeria (Fig. 9.4).

Fig. 9.4 Landsat-8 of Nigeria

Click on the save icon  to save your code.

9.3.3 *Precipitation*

Precipitation data is important for a variety of reasons, including:

- Water resource management: Precipitation data is used to manage water resources such as reservoirs, aquifers, and rivers. This data is critical for planning and implementing water conservation measures, drought management strategies, and flood control measures.
- Agriculture: Precipitation data is used to monitor soil moisture, which is critical for agriculture. Farmers use this data to determine when to plant crops, how much water to irrigate with, and when to harvest.
- Climate studies: Precipitation data is a critical component of climate studies. It helps scientists to understand how weather patterns are changing over time, and how they are affected by various factors such as human activities, natural phenomena, and global warming.
- Public safety: Precipitation data is important for public safety as it can help to predict severe weather conditions such as hurricanes, tornadoes, and floods. This information is used by emergency management officials to issue warnings and evacuation orders and to coordinate response efforts during and after the disaster.

- Transportation: Precipitation data is important for transportation planning and management. It helps to predict road conditions, flight delays, and other travel disruptions caused by weather events.

To visualize the precipitation data using the following code:

```
import ee
from ee_plugin import Map

#Import country boundaries feature collection.
dataset = ee.FeatureCollection('USDOS/LSIB_SIMPLE/2017');

#Apply filter where country name equals Nigeria.
Nigeria = dataset.filter(ee.Filter.eq('country_na', 'Nigeria'));

#Add precipitation data
filtered  =  ee.ImageCollection('UCSB-CHG/CHIRPS/DAILY').filter(ee.
Filter.date('2019-01-01', '2019-01-31')).select('precipitation');
total = filtered.reduce(ee.Reducer.mean());

#Visualization parameters
palette  =  ['1621a2',  'ffffff',  '03ffff',  '13ff03',  'efff00',  'ffb103',
'ff2300'];
visParams = {'min':0, 'max': 10, 'palette': palette};

#Visualizing the data
Map.addLayer(total.clip(Nigeria), visParams,
'NigeriaPrecipitationJan2019');
Map.centerObject(Nigeria, 7);
```

Click run icon ▶, then the result is the precipitation of Nigeria of January 2019 (Fig. 9.5).

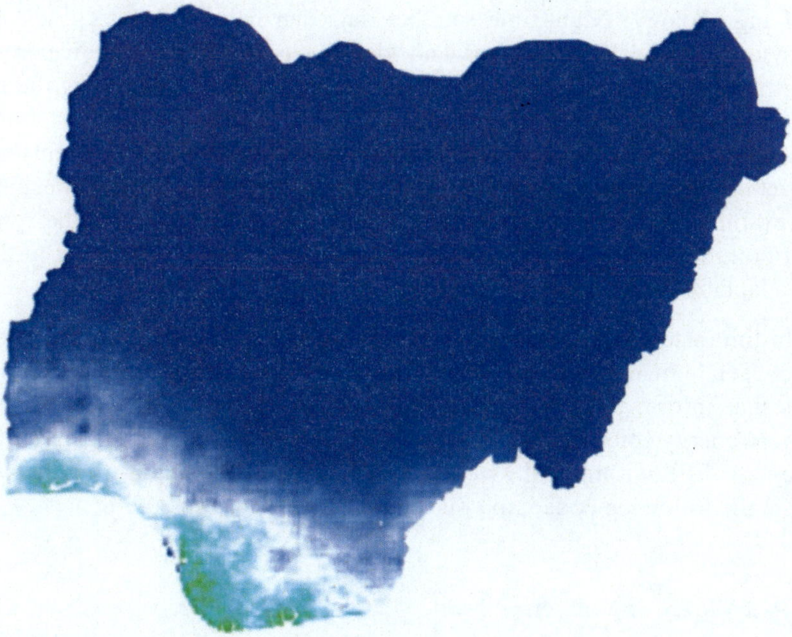

Fig. 9.5 Precipitation of Nigeria

Click on the save icon to save your code.

9.3.4 Night-Time

Night-time satellite data is an important tool for a variety of applications in fields such as climatology, ecology, disaster response, and urban planning. Here are some of the key reasons why night-time satellite data is important:

- Monitoring urbanization: Night-time satellite data can help track the expansion of cities over time, identify areas of high population density, and analyze changes in urbanization patterns. This information is critical for urban planners and policymakers to make informed decisions about land use, transportation, and resource allocation.
- Tracking energy consumption: Night-time satellite data can provide insights into energy consumption patterns by identifying areas with high levels of artificial lighting. This information is important for energy companies and policymakers to develop strategies to reduce energy waste and increase energy efficiency.
- Detecting natural disasters: Night-time satellite data can help detect natural disasters such as fires, floods, and earthquakes and aid in disaster response efforts. For example, it can help emergency responders identify areas with power outages or infrastructure damage and prioritize rescue efforts.

- Studying ecology: Night-time satellite data can help ecologists track animal behavior and migration patterns and identify areas of high biodiversity. This information is important for conservation efforts and understanding the impact of human activities on natural habitats.
- Analyzing climate change: Night-time satellite data can provide insights into the impact of climate change on ecosystems, urban areas, and energy consumption. For example, it can help track changes in urban heat islands and the spread of light pollution, both of which have implications for climate change mitigation and adaptation strategies.

Night-time satellite data is a valuable tool for understanding the complex interactions between human activities, environment, and natural disasters and can provide insights that inform policymaking and decision-making across a wide range of fields from energy consumption to the impact of war, and many things.

Using the GEE within QGIS we can easily visualize the night-time image copy and paste the following code into your python console within QGIS:

```
import ee
from ee_plugin import Map

#Import country boundaries feature collection.
dataset = ee.FeatureCollection('USDOS/LSIB_SIMPLE/2017')

#Apply filter where country name equals Ukraine.
ukraine = dataset.filter(ee.Filter.eq('country_na', 'Ukraine'))

#Defining the datasets
nl2017 = ee.ImageCollection('NOAA/VIIRS/DNB/MONTHLY_V1/VCMSLCFG').
select('avg_rad').filterDate("2017-01-01", "2017-12-31").mean()
nl2022 = ee.ImageCollection('NOAA/VIIRS/DNB/MONTHLY_V1/VCMSLCFG').
select('avg_rad').filterDate("2022-01-01", "2022-12-31").mean()

#Visualization parameters
palette = ['000000', '581845', '900C3F', 'C70039', 'FF5733',
'FFC300', 'DAF7A6']
visParams = {'min':0, 'max': 10, 'palette': palette}

#Visualizing the data
Map.addLayer(nl2017.clip(ukraine), visParams,
'UkraineNighttime2017')
Map.addLayer(nl2022.clip(ukraine), visParams,
'UkraineNighttime2022')
Map.centerObject(ukraine, 7)
```

Then click run icon ▶ it will visualize. The result is Ukraine's night-time image before (2017) and during the Russian invasion (2022) (Fig. 9.6).

Fig. 9.6 Night-time images of Ukraine

Click on the save icon  to save your code.

9.3.5 Copernicus Land Cover

Land use and land cover data (LULC) is an important tool for understanding and managing our natural resources. LULC data provides information about how land is being used and the types of vegetation and other features that cover the surface of Earth. Here are some of the key reasons why LULC data is important:

- Environmental management: Land use and land cover data is critical for environmental management, as it provides a basis for understanding the state of natural resources and for making informed decisions about how to protect and manage them. For example, it can help identify areas of high biodiversity that need to be conserved or areas that are at risk of erosion or desertification.
- Planning and development: LULC data is essential for land use planning and development. By understanding how land is being used and the patterns of development, planners and developers can make informed decisions about where to locate infrastructure, such as roads, buildings, and utilities.
- Natural resource management: LULC data is a valuable tool for managing natural resources, such as forests, water resources, and agricultural land. By understanding how land is being used and the types of vegetation and other features present, resource managers can make informed decisions about how to manage these resources sustainably.
- Climate change mitigation: LULC data is important for monitoring and mitigating the effects of climate change. For example, it can help track changes in vegetation cover, which can affect carbon storage and the absorption of greenhouse gases.

Land use and land cover data is an essential tool for understanding and managing our natural resources. It provides valuable information for environmental management, planning and development, natural resource management, and climate change mitigation.

Using python console to visualize Copernicus data is simple, just copy and paste the following code into your QGIS python console:

```
import ee
from ee_plugin import Map

#Load admin boundary
Nigeria = ee.FeatureCollection('FAO/GAUL/2015/level2').filter(ee.
Filter.inList('ADM0_NAME', ['Nigeria']));

#Load Copernicus data
dataset = ee.Image("COPERNICUS/Landcover/100m/Proba-V-C3/
Global/2019").select('discrete_classification');

#Clip LULC data based on admin boundary
data_clip = dataset.clip(Nigeria);

#Visualize the image
Map.centerObject(Nigeria, 7);
Map.addLayer(data_clip, {}, "Land Cover Nigeria");
```

Click run icon ▶ it will visualize the Copernicus land cover data. The result will be visualized as Fig. 9.7.

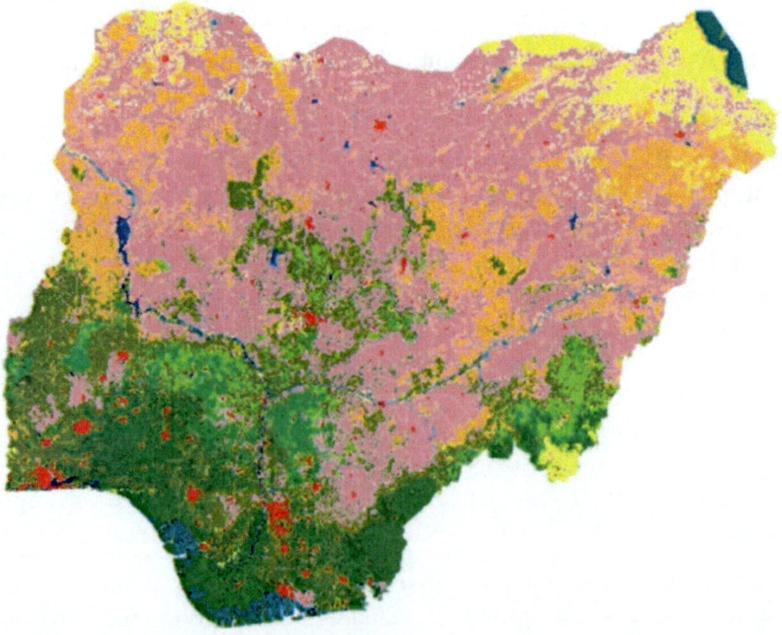

Fig. 9.7 Land cover map of Nigeria

The explanation of each class color is as follows:

More detail about the Copernicus land cover data is available here: https://zenodo.org/record/4723921

9.3.6 ALOS DEM Elevation

Digital Elevation Model (DEM) data is important for a variety of applications in fields such as cartography, geology, engineering, agriculture, urban planning, and environmental management. DEM data represents the elevation of Earth's surface and is typically displayed as a topographic map or a 3D terrain model. Here are some of the key reasons why DEM data is important:

- Land use and land cover analysis: DEM data can be used to determine the slope and aspect of the land, which is important for understanding soil erosion, water flow, and vegetation growth patterns.
- Floodplain mapping: DEM data can be used to map floodplains and to identify areas that are at risk of flooding.
- Resource management: DEM data is used for natural resource management, such as identifying suitable areas for agriculture, forestry, and mining.
- Infrastructure planning: DEM data is essential for infrastructure planning, such as road and bridge construction, pipeline routing, and building siting.
- Military and defense: DEM data is used for military and defense applications, such as terrain analysis, mission planning, and target identification.
- Disaster management: DEM data is crucial for disaster management, such as predicting landslides, assessing earthquake damage, and planning for evacuation routes.

DEM data is a valuable tool for analyzing and understanding Earth's surface, and its importance will only continue to grow as we face complex global challenges such as climate change and population growth.

To visualize the DEM, use the following code. It will open the JAXA ALOS DEM of Madagascar.

```
import ee
from ee_plugin import Map

#Load admin boundary
Madagascar = ee.FeatureCollection('FAO/GAUL/2015/level2').
filter(ee.Filter.inList('ADM0_NAME', ['Madagascar']));

# Load JAXA elevation image.
dataset = ee.Image('JAXA/ALOS/AW3D30/V2_2');
elev = dataset.select('AVE_DSM')

#Clip LULC data based on admin boundary
data_clip = elev.clip(Madagascar);

# Use the terrain algorithms to compute a hillshade with 8-bit values.
shade = ee.Terrain.hillshade(data_clip)
Map.addLayer(shade, {}, 'hillshade')

# Create a "sea" variable to be used for cartographic purposes
sea = data_clip.lte(0)
Map.addLayer(sea.mask(sea), {'palette':'000022'}, 'sea', 0)

# Create a custom elevation palette from hex strings.
elevationPalette = ['006600', '002200', 'fff700', 'ab7634', 'c4d0ff',
'ffffff']
# Use these visualization parameters, customized by location.
visParams = {'min': 1, 'max': 3000, 'palette': elevationPalette}

# Create a mosaic of the sea and the elevation data
visualized = ee.ImageCollection([
  # Mask the elevation to get only land
  data_clip.mask(sea.Not()).visualize(**visParams),
  # Use the sea mask directly to display sea.
  sea.mask(sea).visualize(**{'palette':'000022'})
]).mosaic()

# Note that the visualization image doesn't require visualization
parameters.
Map.addLayer(visualized, {}, 'elevation palette')

# Convert the visualized elevation to HSV, first converting to [0,
1] data.
```

```
hsv = visualized.divide(255).rgbToHsv()
# Select only the hue and saturation bands.
hs = hsv.select(0, 1)
# Convert the hillshade to [0, 1] data, as expected by the HSV
algorithm.
v = shade.divide(255)
# Create a visualization image by converting back to RGB from HSV.
# Note the cast to byte in order to export the image correctly.
rgb = hs.addBands(v).hsvToRgb().multiply(255).byte()
Map.addLayer(rgb, {}, 'styled')

# Zoom to an area of interest.
Map.centerObject(Madagascar, 6);
```

Then click run icon, the result will produce three different layers: styled, elevation, and hillshade (Fig. 9.8).

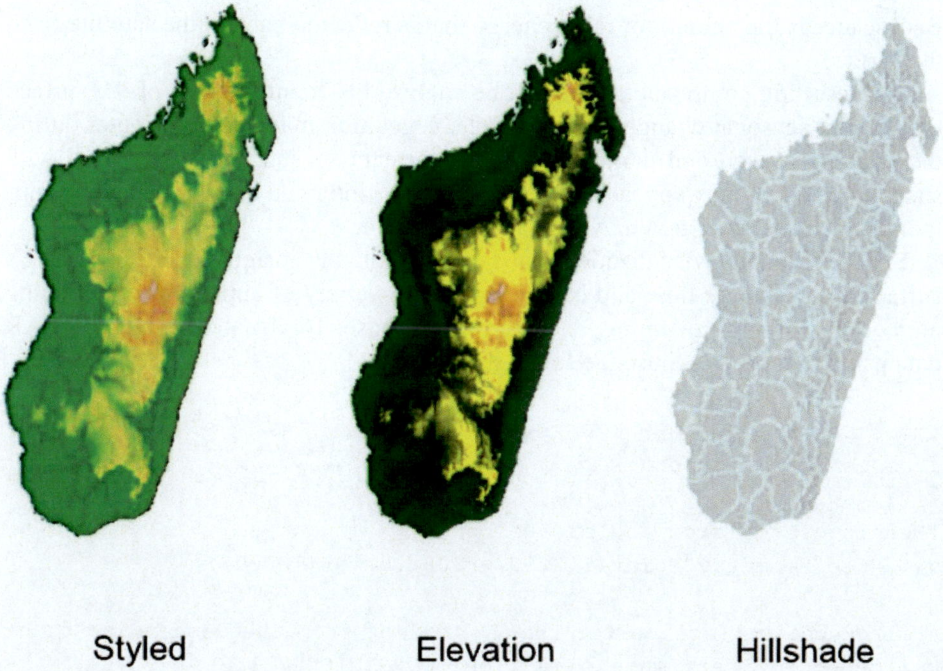

Fig. 9.8 Different visualization of elevation

Do not forget to click on the save icon to save your code.

9.3.7 Sentinel-1 (SAR Seasonality)

Sentinel-1 is a satellite mission developed by the European Space Agency (ESA) for Earth's observation. One of the primary sensors onboard the Sentinel-1 satellites is the Synthetic Aperture Radar (SAR), which is capable of collecting high-resolution images of Earth's surface, regardless of weather conditions or the time of day.

Sentinel-1 SAR data can be used to analyze changes in Earth's surface over time. One important aspect of this analysis is understanding how the surface changes seasonally. SAR seasonality visualization is a technique used to visualize these changes.

SAR seasonality visualization involves collecting multiple SAR images of the same area over a period of time (usually 1 year or more) and then comparing them. By comparing the images, it is possible to identify changes in the surface that occur over different seasons.

The visualization process typically involves stacking the SAR images on top of each other to create a single composite image. Each pixel in the composite image represents the average backscatter value for that pixel over the entire time period. Backscatter is the amount of radar energy that is reflected back to the satellite from the surface.

The resulting composite image can be analyzed to identify areas of the surface that exhibit seasonal changes. For example, vegetation may appear brighter during the growing season and darker during the dormant season. Similarly, areas with snow or ice cover may appear brighter during the winter and darker during the summer. While urbanized area may appear bright white.

SAR seasonality visualization can provide valuable insights into how Earth's surface changes over time and can be used for a variety of applications, including agriculture, land use planning, and disaster response. To visualize Sentinel-2 SAR data using python QGIS use the following code:

```
import ee
from ee_plugin import Map

#Defining study area and create buffer
roi = ee.Geometry.Point(112.59,-8.09).buffer(10000);

#Filter the collection for the VV product from the descending track
collectionVV = ee.ImageCollection('COPERNICUS/S1_GRD')\
    .filter(ee.Filter.eq('instrumentMode', 'IW'))\
    .filter(ee.Filter.listContains('transmitterReceiverPolarisat
ion', 'VV'))\
    .filter(ee.Filter.eq('orbitProperties_pass', 'DESCENDING'))\
    .filterBounds(roi)\
    .select(['VV']);
print(collectionVV);
```

```
#Filter the collection for the VH product from the descending track
collectionVH = ee.ImageCollection('COPERNICUS/S1_GRD')\
    .filter(ee.Filter.eq('instrumentMode', 'IW'))\
    .filter(ee.Filter.listContains('transmitterReceiverPolarisat
ion', 'VH'))\
    .filter(ee.Filter.eq('orbitProperties_pass', 'DESCENDING'))\
    .filterBounds(roi)\
    .select(['VH']);
print(collectionVH);

#Let's centre the map view over our ROI
Map.centerObject(roi, 13);

#Create a 3 band stack by selecting from different periods (months)
VV1 = ee.Image(collectionVV.filterDate('2022-01-01', '2022-04-30').
median());
VV2 = ee.Image(collectionVV.filterDate('2022-05-01', '2022-08-31').
median());
VV3 = ee.Image(collectionVV.filterDate('2022-09-01', '2022-12-31').
median());

composite = VV1.addBands(VV2).addBands(VV3)


#Add to map
Map.addLayer(composite.clip(roi), {'min': -12, 'max': -7}, 'Season
composite');
```

The seasonality of sugarcane plantation of Malang, East Java, Indonesia, in the center of ROI is clearly shown as a result (Fig. 9.9).

Fig. 9.9 Seasonality map of SAR images

9.3.8 Sentinel-5 (NO2)

Sentinel-5 is a satellite mission developed by the European Space Agency (ESA) that is part of the Copernicus Program. It is designed to monitor Earth's atmosphere and provide accurate information on the composition of the air we breathe, including pollutants, greenhouse gases, and other atmospheric trace gases.

The Sentinel-5 data is important because it provides valuable information for understanding and monitoring Earth's environment and climate. Here are some key reasons why Sentinel-5 data is significant:

- Monitoring air quality: Sentinel-5 data can be used to monitor air quality by providing information on pollutants such as nitrogen dioxide, sulfur dioxide, carbon monoxide, and ozone. This information can help identify areas with poor air quality and support efforts to reduce pollution levels.
- Understanding climate change: Sentinel-5 data can be used to track the concentrations of greenhouse gases, such as carbon dioxide and methane, in the atmosphere. This data can help scientists understand how human activities are contributing to climate change and support efforts to mitigate its effects.
- Assessing health risks: Sentinel-5 data can be used to assess the impact of air pollution on human health. For example, exposure to high levels of nitrogen dioxide can cause respiratory problems, and the data can be used to identify areas where people are at risk of exposure.

- Supporting disaster response: Sentinel-5 data can be used to monitor the spread of smoke from wildfires, volcanic eruptions, and other natural disasters. This information can be used to assess the risks to human health and support disaster response efforts.

Sentinel-5 data plays a critical role in helping scientists and policymakers make informed decisions about environmental and climate-related issues. It provides accurate and timely information that is essential for protecting our planet and its inhabitants. In this case we will visualize NO2 datasets of Jakarta Megacity before and during COVID19, using the following code:

```
import ee
from ee_plugin import Map

#Defining study area and create buffer
point = ee.Geometry.Point(106.85, -6.25).buffer(100000);

#Defining datasets and time range
before = ee.ImageCollection('COPERNICUS/S5P/NRTI/L3_NO2')\
  .select('NO2_column_number_density')\
  .filterDate('2019-12-01', '2019-12-31');

after = ee.ImageCollection('COPERNICUS/S5P/NRTI/L3_NO2')\
  .select('NO2_column_number_density')\
  .filterDate('2020-04-01', '2020-04-30');

#Visualization parameters
vizParams = {'min': 0, 'max': 0.0002, \
           'palette':['black', 'blue', 'purple', 'cyan', 'green',
'yellow', 'red'],\
            'opacity': 0.7}

#Display the result
Map.addLayer(before.mean().clip(point),        vizParams,      'Before
COVID19');
Map.addLayer(after.mean().clip(point),         vizParams,       'After
COVID19');
Map.setCenter(106.85, -6.25, 9);
```

The result will be shown as Fig. 9.10, while details about the study have been published by Ramdani and Setiani (2022).

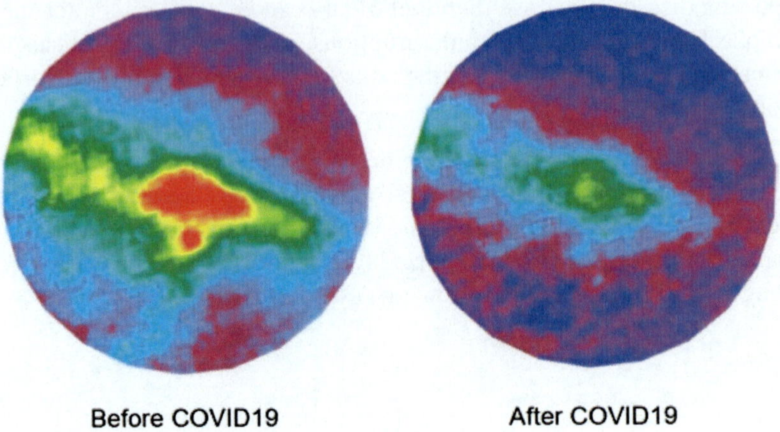

Fig. 9.10 SO2 maps before and after COVID-19 pandemic

9.3.9 Land Surface Temperature

The Moderate Resolution Imaging Spectroradiometer (MODIS) is an instrument onboard NASA's Terra and Aqua satellites that provides important data on various aspects of Earth's surface, including land surface temperature (LST).

LST is a key parameter for understanding many physical, biological, and environmental processes that occur at Earth's surface. Some of the main reasons why LST is important include:

- Climate modeling: LST is an important input parameter for climate models used to understand and predict Earth's climate. By accurately measuring LST, scientists can improve the accuracy of climate models and their ability to forecast climate changes.
- Land cover and land use analysis: LST can be used to monitor changes in land cover and land use, which are critical for understanding the impact of human activities on the environment. For example, LST can help identify urban heat islands and monitor changes in agricultural lands.
- Monitoring natural disasters: LST can be used to monitor natural disasters such as wildfires, droughts, and floods. By detecting changes in LST, scientists can quickly identify areas at risk and take necessary measures to minimize the damage caused by these events.
- Monitoring air pollution: LST can be used to monitor air pollution, which is a major environmental and health concern. By measuring LST, scientists can track the movement of pollutants and their effects on the environment and public health.

MODIS provides high-quality data on LST that is used by scientists and policymakers around the world to address a range of environmental challenges. To visualize the LST of MODIS using python console within QGIS you can use the following code:

```
import ee
from ee_plugin import Map

#Defining study area
Madagascar = ee.FeatureCollection('FAO/GAUL/2015/level2').
filter(ee.Filter.inList('ADM0_NAME', ['Madagascar']));

#Defining datasets
dataset = ee.ImageCollection('MODIS/061/MOD11A1').filter(ee.
Filter.date('2019-01-01', '2019-01-31'));
modis = dataset.select('LST_Day_1km');
total = modis.reduce(ee.Reducer.mean());
lst = total.clip(Madagascar)

#Visualization parameters
palette = ['040274', '040281', '0502a3', '0502b8', '0502ce',
    '0502e6', '0602ff', '235cb1', '307ef3', '269db1', '30c8e2',
    '32d3ef', '3be285', '3ff38f', '86e26f', '3ae237', 'b5e22e',
    'd6e21f', 'fff705', 'ffd611', 'ffb613', 'ff8b13', 'ff6e08', 'ff500d',
    'ff0000', 'de0101', 'c21301', 'a71001', '911003']
visParamsK = {'min':14000, 'max': 16000, 'palette': palette}

#Convert to Celsius
modisc = lst.multiply(0.02).subtract(273.15)

#Visualization parameters
palette = ['blue', 'limegreen', 'yellow', 'darkorange', 'red']
visParamsC = {'min': 20, 'max': 40, 'palette': palette}

# Zoom to an area of interest.
Map.centerObject(Madagascar, 7);

#Add clipped image layer to the map.
Map.addLayer(lst, visParamsK, 'Mean temperature of Jan 2019
(Kelvin)');
Map.addLayer(modisc, visParamsC, 'Mean temperature of
Jan 2019 (Celsius)').
```

Then click run icon ▶ it will visualize the LST of Madagascar on January 2019 in Kelvin and Celsius. The result will be as Fig. 9.11.

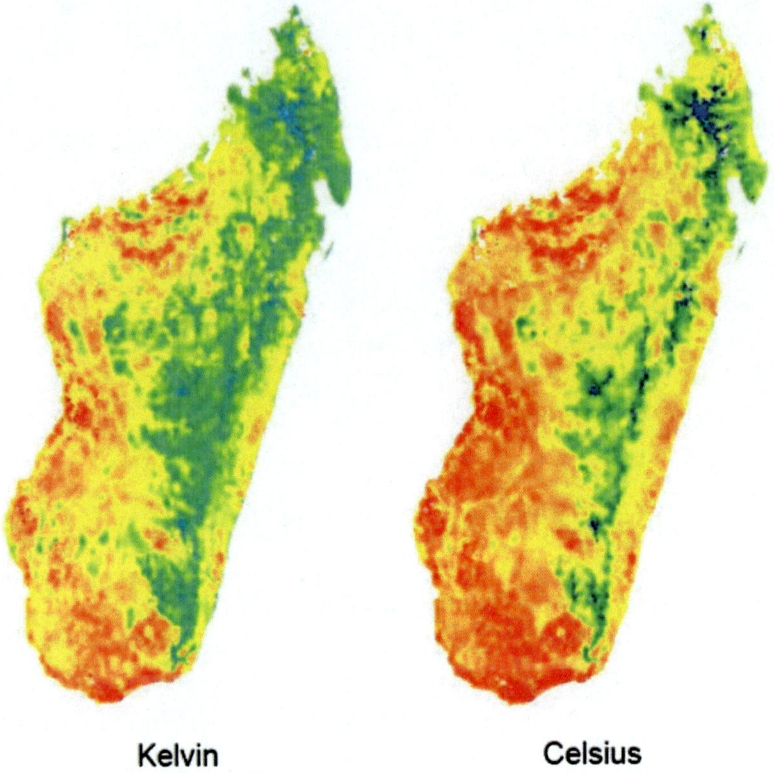

Fig. 9.11 LST of Madagascar in Kelvin and Celsius unit

Always remember to click on the save icon to save your code.

9.3.10 Sea Surface Temperature

Sea surface temperature (SST) data is a crucial component of understanding Earth's climate system. Here are some reasons why SST data is important:

- Climate modeling: SST data is a key input parameter for climate models. These models are used to predict future climate conditions and to understand how the climate system has changed in the past. By accurately simulating SST, climate models can make more accurate predictions about the future state of the planet's climate.
- Understanding ocean circulation: SST data can help us understand ocean circulation patterns, which play a critical role in regulating global climate. The temperature of the surface layer of the ocean affects the density of water, which in turn affects ocean currents. By monitoring SST, scientists can better understand how ocean currents are changing and how this might impact global climate.

- Marine ecosystems: SST data is used to monitor changes in the ocean's temperature that can affect marine ecosystems. For example, changes in SST can affect the distribution of marine species, which can have significant impacts on the fishing industry and the overall health of ocean ecosystems.
- Weather forecasting: SST data is used to improve weather forecasting. By monitoring changes in SST, meteorologists can better predict the development and movement of storms, which can help to save lives and protect property.

SST data is an important tool for understanding and predicting changes in Earth's climate system. It helps scientists and policymakers to make informed decisions about how to address climate change and protect the planet's natural resources.

Open your python console within QGIS and paste the following code to visualize the SST:

```
import ee
from ee_plugin import Map

#Defining datasets
dataset = ee.ImageCollection("JAXA/GCOM-C/L3/OCEAN/SST/V3").fil-
terDate('2022-12-01', '2023-01-01').filter(ee.Filter.
eq("SATELLITE_DIRECTION", "D")); # filter to daytime data only

#Multiply with slope coefficient and add offset
dataset = dataset.mean().multiply(0.0012).add(-10);

#Visualization parameters
palette = ['000000', '005aff', '43c8c8', 'fff700', 'ff0000']
visParams = {'bands': 'SST_AVE', 'min':0, 'max':30, 'palette':
palette}

#Map and zoom to specific area
Map.setCenter(128.45, 33.33, 5);
Map.addLayer(dataset, visParams, "Sea Surface Temperature");
```

The result is as follows. It is visualized the SST around East Asia region on early December 2022 to early January 2023 (Fig. 9.12).

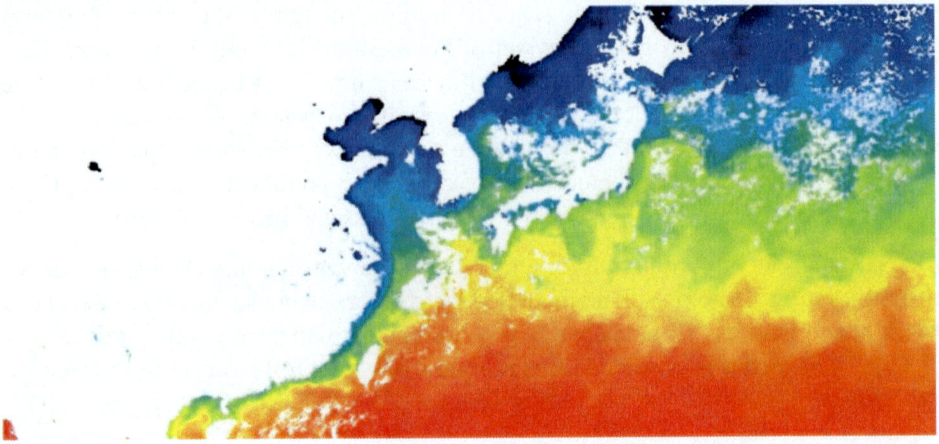

Fig. 9.12 SST around east Asia region

Click on the save icon to save your code.

9.4 Map Algebra (Expression)

The map algebra is a mathematical approach used to process and analyze geospatial data in GIS software. It involves the use of mathematical operations and functions to manipulate and combine different raster layers or grids of data, such as elevation, land use, precipitation, temperature, and so on.

Using expression in Python QGIS enables users to perform map algebra operations on raster layers or grids of data. This approach allows for complex and sophisticated analyses of spatial data, such as overlaying multiple layers to create a composite map, calculating terrain features such as slope and aspect, or performing suitability analysis for a particular activity. Some of the benefits of using map algebra or expressions in Python QGIS include:

- Enhanced analysis: Map algebra allows users to perform complex and sophisticated analysis on raster layers or grids of data that are not possible with simple GIS tools. This enables better decision-making and planning.
- Increased efficiency: Map algebra or expressions in Python QGIS automate repetitive tasks, saving time and effort for users.
- Greater flexibility: Map algebra allows users to combine data from different sources and analyze them using various mathematical operations and functions, providing greater flexibility in analysis.
- Improved accuracy: Map algebra allows users to perform more precise analysis, which can improve the accuracy of the results.

Using python console within QGIS are essential tools for spatial data analysis, enabling users to perform complex and sophisticated analysis on raster layers or grids of data, enhancing efficiency and flexibility, and improving accuracy.

9.4.1 SAVI and NDVI

In this case, we will calculate the Soil-Adjusted Vegetation Index (SAVI) and Normalized Difference Vegetation Index (NDVI) of Landsat-8. SAVI and NDVI are two commonly used vegetation indices that provide valuable information about vegetation cover, health, and productivity. Both indices are calculated from the same satellite data but use different algorithms to emphasize different aspects of vegetation.

The NDVI is a measure of the amount of healthy vegetation in an area, based on the difference between the amount of red and near-infrared light reflected by vegetation. NDVI values range from −1 to +1, with higher values indicating more healthy and dense vegetation. NDVI is used to monitor vegetation growth, track changes in land cover, and assess crop health.

The SAVI index, on the other hand, adjusts for the influence of soil brightness on vegetation reflectance. Soil brightness can sometimes interfere with NDVI readings, particularly in arid areas where bare soil can have a strong influence on the reflectance values. SAVI attempts to correct this by adjusting the NDVI values based on the amount of bare soil present in the area. This makes SAVI a more accurate measure of vegetation health in areas with significant soil background effects.

Both NDVI and SAVI are important tools for monitoring vegetation cover and health, which are critical for a range of applications including agriculture, forestry, conservation, and climate change research. By using these indices, scientists and policymakers can better understand the dynamics of vegetation growth and change over time, which can inform decisions about land use, resource management, and environmental policy.

Copy and paste the following code into your python QGIS console:

```
import ee
from ee_plugin import Map

#Defining study area
Madagascar    =    ee.FeatureCollection('FAO/GAUL/2015/level2').
filter(ee.Filter.inList('ADM0_NAME', ['Madagascar']));

#Filter the L8 collection to a single month.
collection = ee.ImageCollection('LANDSAT/LC08/C02/T1_RT_TOA') \
    .filterDate('2022-11-01', '2022-12-01')

#Map an expression over a collection.
```

```
#A function to compute NDVI.
def NDVI(image):
   return image.expression('float(b("B5") - b("B4")) / (b("B5") +
b("B4"))')

#A function to compute SAVI.
def SAVI(image):
  return image.expression(
      '(1 + L) * float(nir - red)/ (nir + red + L)',
      {
        'nir': image.select('B5'),
        'red': image.select('B4'),
        'L': 0.2
      })

#Visualization parameters.
vis = {
  'min': 0,
  'max': 1,
  'palette': [
      'FFFFFF', 'CE7E45', 'DF923D', 'F1B555', 'FCD163', '99B718',
      '74A901', '66A000', '529400', '3E8601', '207401', '056201',
      '004C00', '023B01', '012E01', '011D01', '011301'
  ]
}

#Zoom to an area of interest.
Map.centerObject(Madagascar, 7);

#Computes the mean NDVI and SAVI and clip to the study area.
meanNDVI = collection.map(NDVI).mean().clip(Madagascar)
meanSAVI = collection.map(SAVI).mean().clip(Madagascar)

#Map and display.
Map.addLayer(meanNDVI, vis, 'Mean NDVI')
Map.addLayer(meanSAVI, vis, 'Mean SAVI')
```

Click run icon ▶ it will visualize the SAVI and NDVI of Madagascar at the end of year 2022. Then click the save icon to save your code. The result will be visualized as Fig. 9.13.

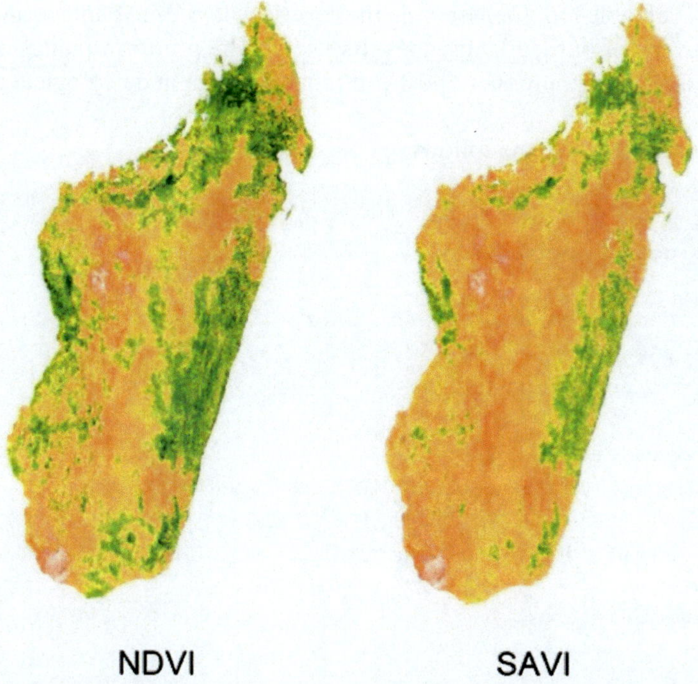

Fig. 9.13 Map of NDVI and SAVI

All codes used in this chapter are available at: https://github.com/fatwaramdani/QGIS-GEE.

9.4.2 Normalized Burn Ratio

Normalized Burn Ratio (NBR) is an index that is commonly used in remote sensing to assess the severity of burn scars in vegetation after a wildfire. The NBR measures the difference between near-infrared (NIR) and shortwave infrared (SWIR) reflectance, which is sensitive to changes in vegetation structure and moisture content.

The importance of the NBR lies in its ability to provide valuable information about the extent and severity of a wildfire. By analyzing the NBR of pre- and post-fire imagery, researchers can identify areas of vegetation that have been severely burned and areas that have not been affected. This information can be used to monitor the recovery of the burned area over time and to assess the ecological impacts of the fire.

The NBR is also useful in predicting post-fire erosion and runoff, as vegetation loss can increase the likelihood of soil erosion and flash floods. Additionally, the NBR can be used to evaluate the effectiveness of post-fire rehabilitation efforts, such as reseeding or mulching, by comparing pre- and post-treatment NBR values.

NBR is a valuable tool for assessing the severity of wildfire damage and monitoring the recovery of burned landscapes. Its use can help inform management decisions related to fire suppression, post-fire rehabilitation, and ecological restoration efforts.

To calculate NBR use the following code:

```
import ee
from ee_plugin import Map

#Defining study area and create buffer
point = ee.Geometry.Point(112.59, -7.75).buffer(10000);
Map.centerObject(point, 11);

#Defining datasets
s2 = ee.ImageCollection('COPERNICUS/S2_HARMONIZED')\
    .select(['B2', 'B3', 'B4', 'B8', 'B11', 'B12'],\
        ['blue', 'green', 'red', 'nir', 'swir1', 'swir2']);

#Before forest fire
preImage = s2\
    .filterBounds(point)\
    .filterDate('2019-06-20', '2019-06-30')\
    .sort('CLOUD_COVERAGE_ASSESSMENT')\
    .first();

#After forest fire
postImage = s2\
    .filterBounds(point)\
    .filterDate('2019-10-20', '2019-10-31')\
    .sort('CLOUD_COVERAGE_ASSESSMENT')\
    .first();

#Vizualization parameters for RGB combinations
visParam = {\
    'bands': ['swir2', 'swir1', 'green'],\
    'min': 0,\
    'max': 3000\
};

Map.addLayer(preImage.clip(point), visParam, 'pre');
Map.addLayer(postImage.clip(point), visParam, 'post');

#Calculate NBR.
nbrPre = preImage.normalizedDifference(['nir', 'swir2'])\
    .rename('nbr_pre');
nbrPost = postImage.normalizedDifference(['nir', 'swir2'])\
    .rename('nbr_post');
```

```
#Calculate change.
diff = nbrPost.subtract(nbrPre).rename('change');

palette = [\
    '011959', '0E365E', '1D5561', '3E6C55', '687B3E',\
    '9B882E', 'D59448', 'F9A380', 'FDB7BD', 'FACCFA'\
];

#Vizualization parameters for dNBR
visParams = {\
    'palette': palette,\
    'min': -1,\
    'max': 1\
};

#Display the result, clip to study area
Map.addLayer(diff.clip(point), visParams, 'change');

#Classify change
thresholdGain = 0.5;
thresholdLoss = -0.50;

diffClassified = ee.Image(0);


diffClassified = diffClassified.where(diff.lte(thresholdLoss), 2);
diffClassified = diffClassified.where(diff.gte(thresholdGain), 1);

#Vizualization parameters for threshold
changeVis = {\
    'palette': ['fff70b', 'ffaf38', 'ff641b'],\
   'min': 0,\
   'max': 2\
};

#Display the final result
Map.addLayer(diffClassified.selfMask().clip(point),      changeVis,
'change classified by threshold');
```

The code will produce four different images: pre-event, post-event, change, and change classified by threshold (Fig. 9.14). Details about the study has been published by Setiani, P., Devianto, L.A. & Ramdani, F. Rapid estimation of CO2 emissions from forest fire events using cloud-based computation of Google Earth Engine. Environ Monit Assess 193, 669 (2021). https://doi.org/10.1007/s10661-021-09460-w

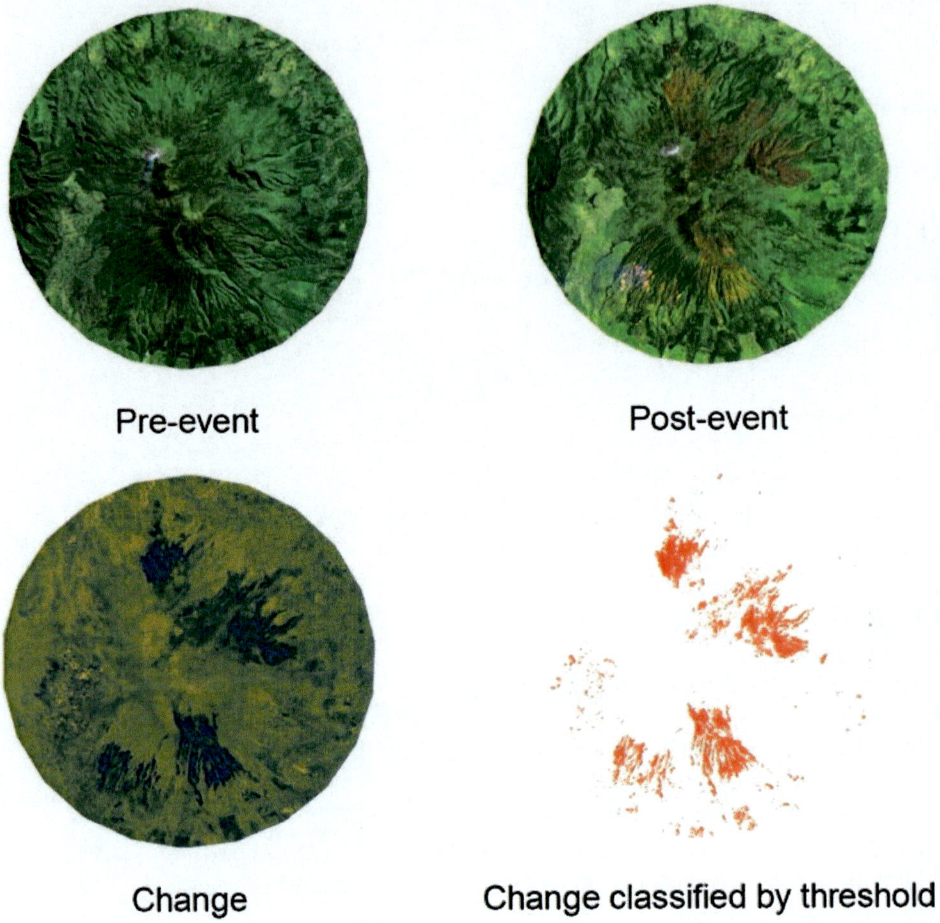

Fig. 9.14 Forest fire events

For modification purposes, Table 9.1 guides you on changing the parameters.

Table 9.1 Parameters configuration of GEE

Parameters	Description
Study area	Change the country name of your study area, for example: `['Madagascar'] >> ['Kenya']` Change the coordinate and buffer size, for example: `point = ee.Geometry.Point(112.59, -7.75).buffer(10000) >> point = ee.Geometry.Point(106.85, -6.25).buffer(5000)`
Datasets	Change the data collection, for example: `ee.ImageCollection('LANDSAT/LC08/C02/T1_RT_TOA') >> ee.ImageCollection('LANDSAT/LE07/C02/T1')`
Date	Change the date range, for example: `.filterDate('2022-11-01', '2022-12-01') >> .filterDate('2023-01-01', '2023-02-01')`
Visualization	Change the parameter of min, max, and palette. Especially for palette, you can directly use "red." "blue," or "green" instead of color code in HTML
Zoom level	Change the number, for example: `Map.centerObject(Madagascar, 7) >> Map.centerObject(Madagascar, 9);`

Chapter 10
Case Studies and Examples of Earth Remote Sensing Applications Using QGIS

This chapter will cover some case studies and examples of how QGIS can do to unlock the potential of satellite imageries.

In this comprehensive chapter, we will delve into a fascinating array of compelling case studies and real-world examples, showcasing the remarkable capabilities of QGIS in harnessing and unleashing the immense potential of satellite imagery. By exploring a diverse range of scenarios and applications, we will demonstrate how QGIS empowers users to unlock invaluable insights, unravel complex patterns, and extract meaningful information from an extensive repertoire of satellite images.

10.1 Vegetation Cover Change in New Delhi

Vegetation cover provides many positive things for urban inhabitants, from fresh air to lower surface temperature. In this case study we will quantify and map the spatial distribution of vegetation cover in New Delhi City, India.

The first thing we need is the city level boundary. We will extract from Natural Earth data and save it. Select New Delhi using the "Select feature" icon  then click Layer > Save as (Fig. 10.1).

F. Ramdani, *Exploring the Earth with QGIS*, Springer Remote Sensing/Photogrammetry, https://doi.org/10.1007/978-3-031-46042-5_10

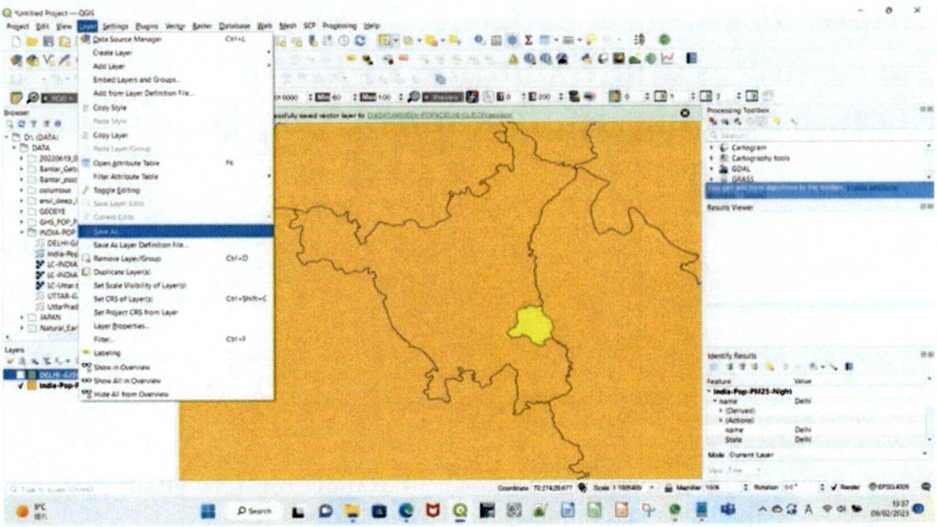

Fig. 10.1 Select only New Delhi

You need to check the "Save only selected features" so only the New Delhi boundary will be saved. In the "Format" select GeoJSON (Fig. 10.2). We will use it to download the satellite data.

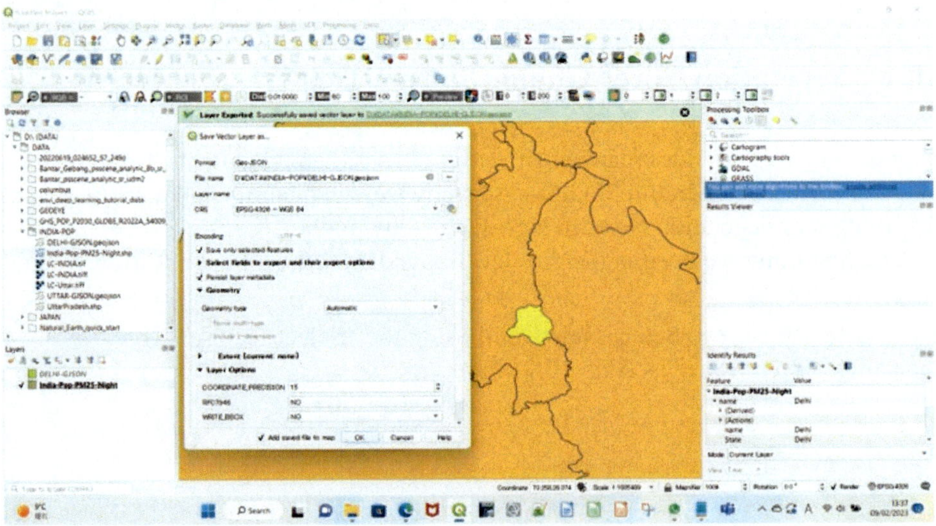

Fig. 10.2 Save as GeoJSON

Next, open the browser and type https://dataspace.copernicus.eu/, then click Login button. The new window will open and type "Delhi" in the search window. Then upload the GeoJSON data that has saved before by clicking on the "Create an area of interest" icon . Click on the "Upload a file to create" icon (Fig. 10.3).

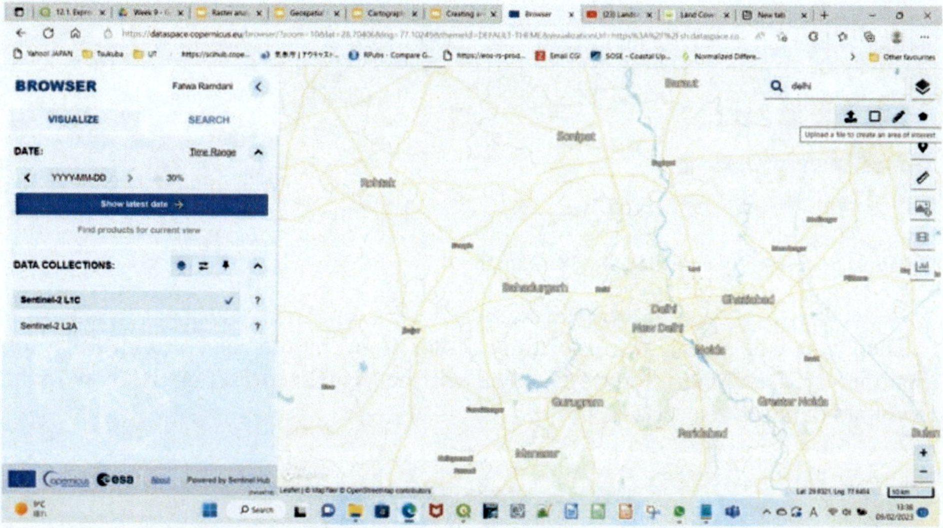

Fig. 10.3 Upload an area of interest

Select your "New Delhi" GeoJSON data (Fig. 10.4).

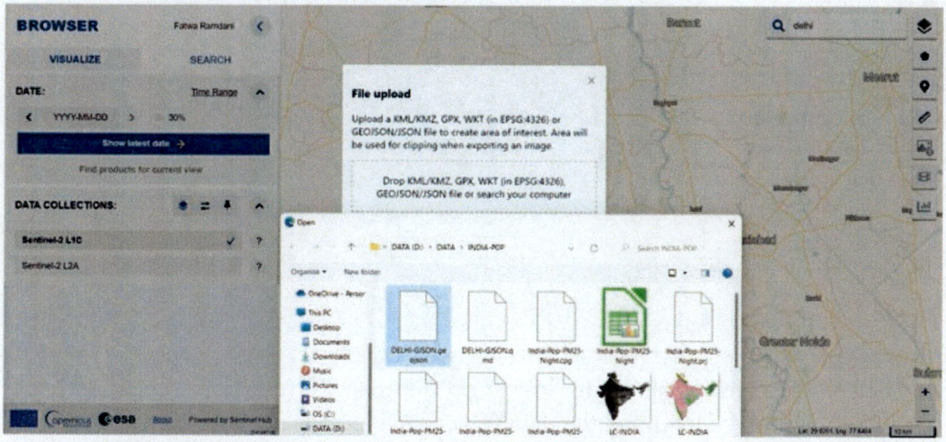

Fig. 10.4 Select the GeoJSON file to be uploaded

If your data successfully uploaded, it will be visualized like Fig. 10.5.

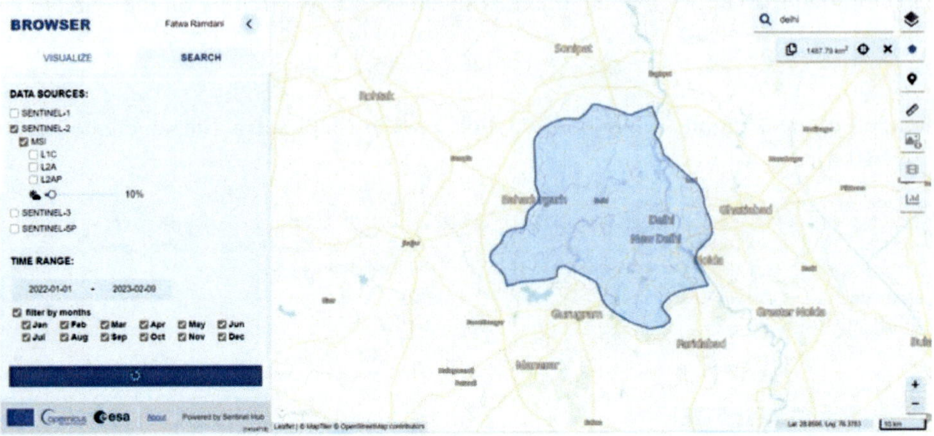

Fig. 10.5 Successfully uploaded GeoJSON data

Then you can select the time range, filter it by month, and search only for "Sentinel-2" data. The filtered data then will be visualized as Fig. 10.6. We will download data for year 2023.

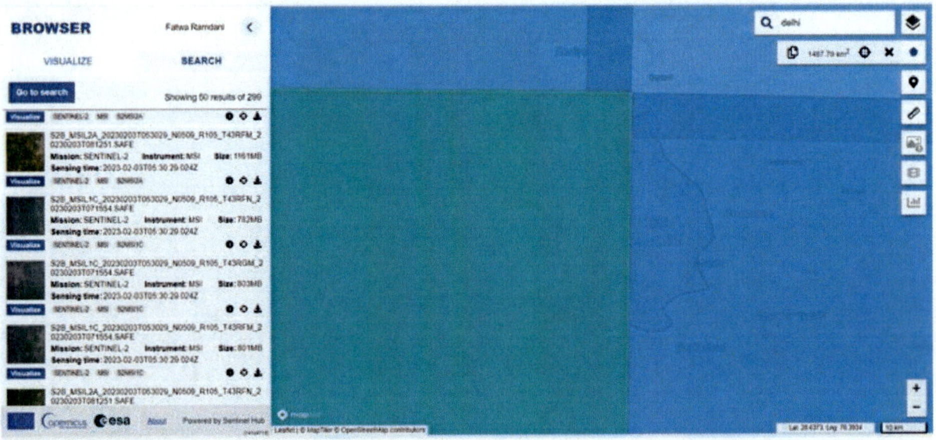

Fig. 10.6 Result of filtered data

Click "Visualize" to show the image on the map. If you are satisfied with the result then you may download the data. You can also create different type of visualization, such as "True Color," "False Color," and etc. (Fig. 10.7).

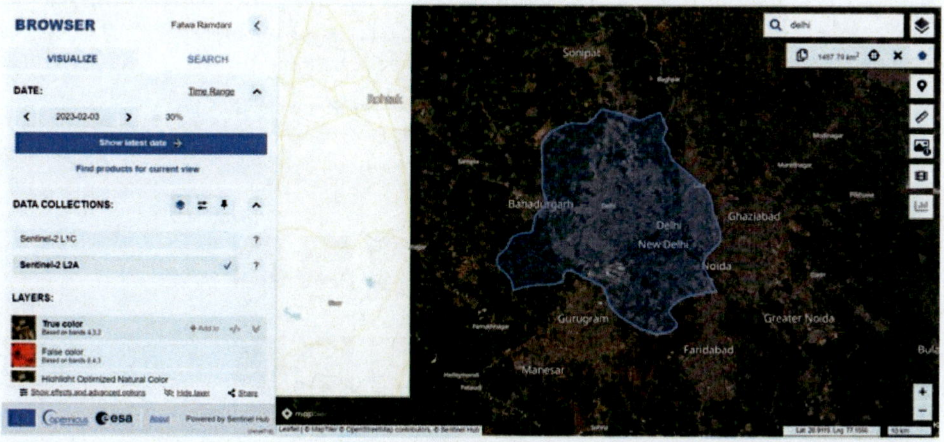

Fig. 10.7 Select different visualization

To download, click on the "Download Image" icon . A new window will appear and select "Analytical." Change the image format to TIFF (8-bit), image resolution to MEDIUM, coordinate system to WGS 84 (EPSG: 4326), and select all "Raw" data. Click "Download" (Fig. 10.8).

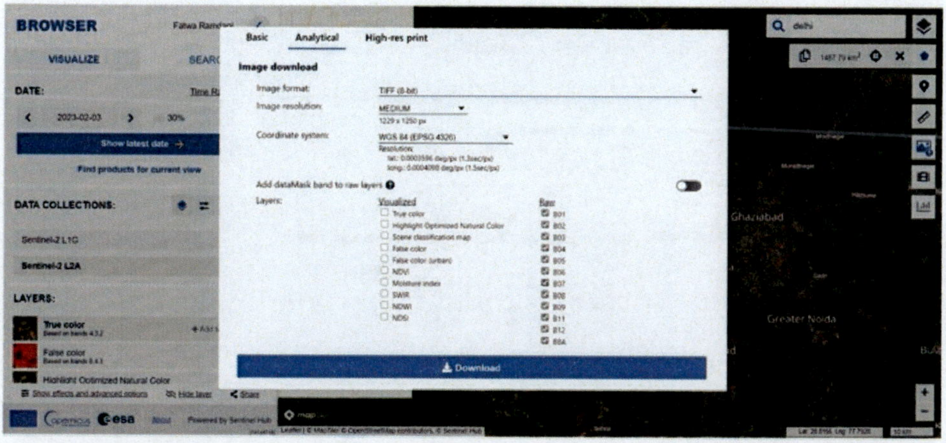

Fig. 10.8 Download analytical data

Extract all your downloaded data then open QGIS. Make sure the SCP plugin is already installed and then click on the icon  to activate. Click on "Band set" then click on "Open file" icon , select all bands (Fig. 10.9).

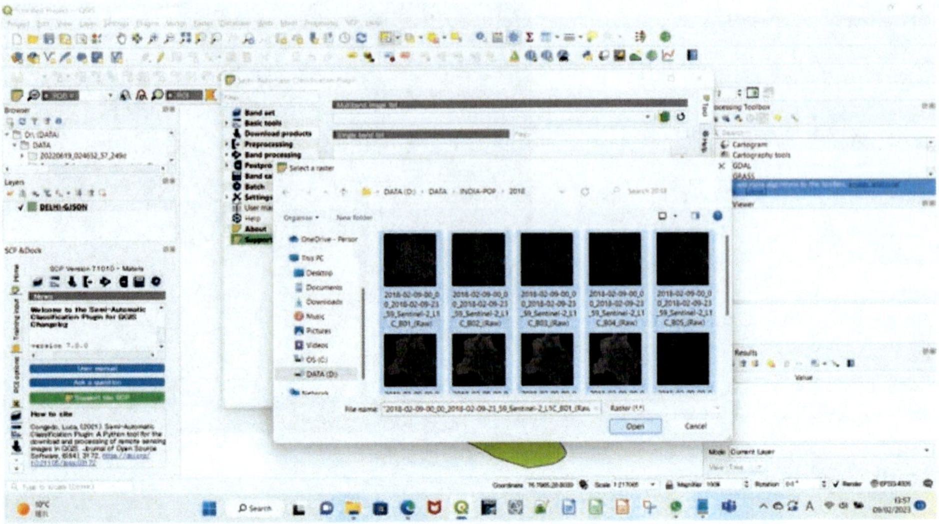

Fig. 10.9 Open downloaded data within QGIS using SCP plugin

Check on "Create virtual raster" and "Create raster of band" (Stack bands) then click RUN (Fig. 10.10).

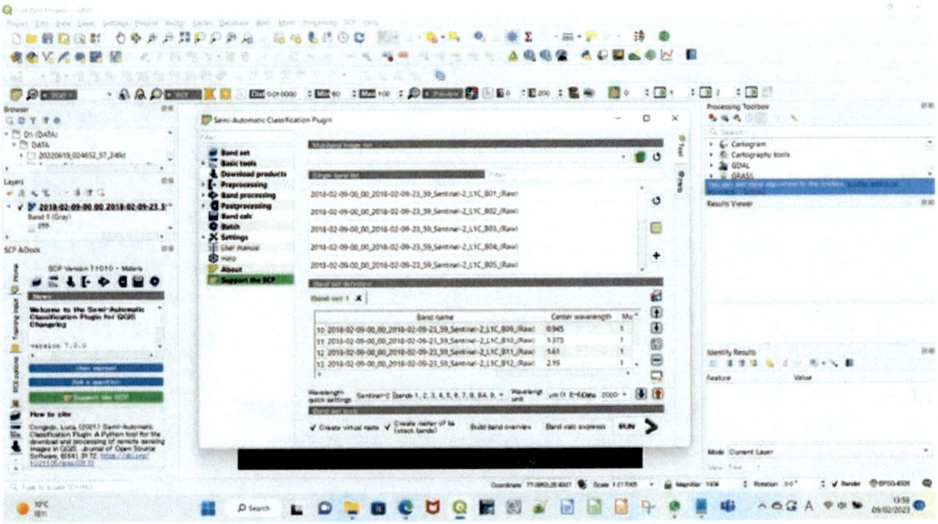

Fig. 10.10 Create band set

Wait for couple seconds, then in the main canvas will appear the result as shown in Fig. 10.11.

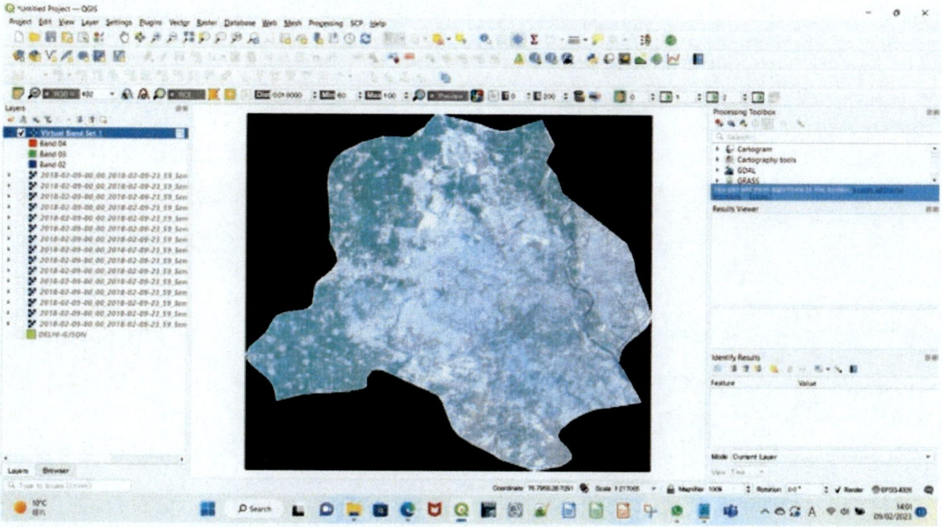

Fig. 10.11 Result of RGB band set

You can change the visualization by double-clicking on the "Virtual Band Set" layer and change the Red band to "Band 04," Green band to "Band 03," and Blue band to "Band 02" in the layer properties dialog box (Fig. 10.12).

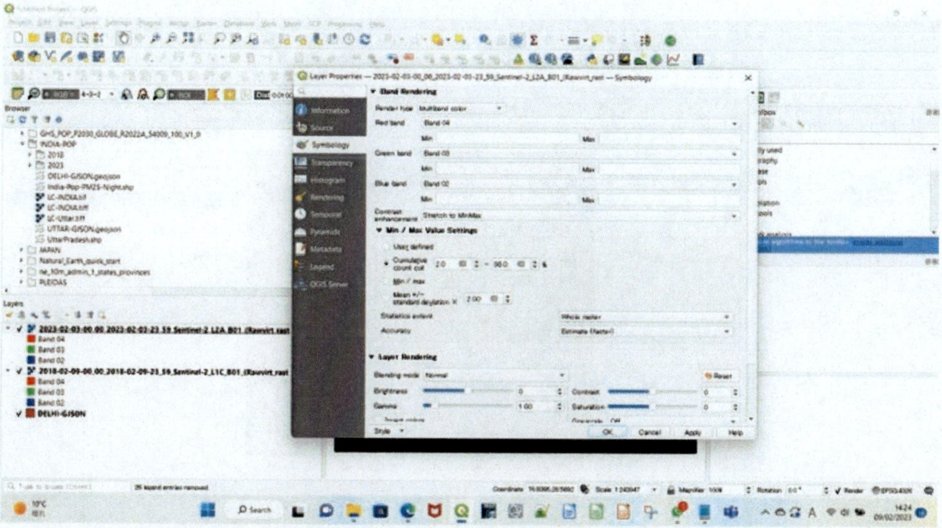

Fig. 10.12 Change the RGB combination

Or just input directly on the "Show/hide input image" icon 

. Just type 432 then hit Enter. The result will be shown as Fig. 10.13.

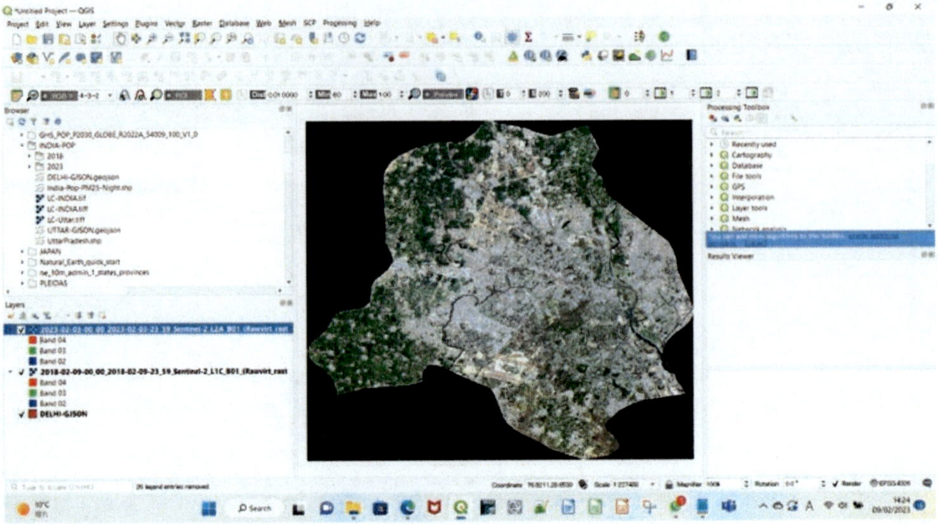

Fig. 10.13 RGB 432 combination

Next, we can use "Raster Calculator" to calculate the vegetation index, which means we will get the density of vegetation from the input raster. Click on Raster > Raster Calculator (Fig. 10.14).

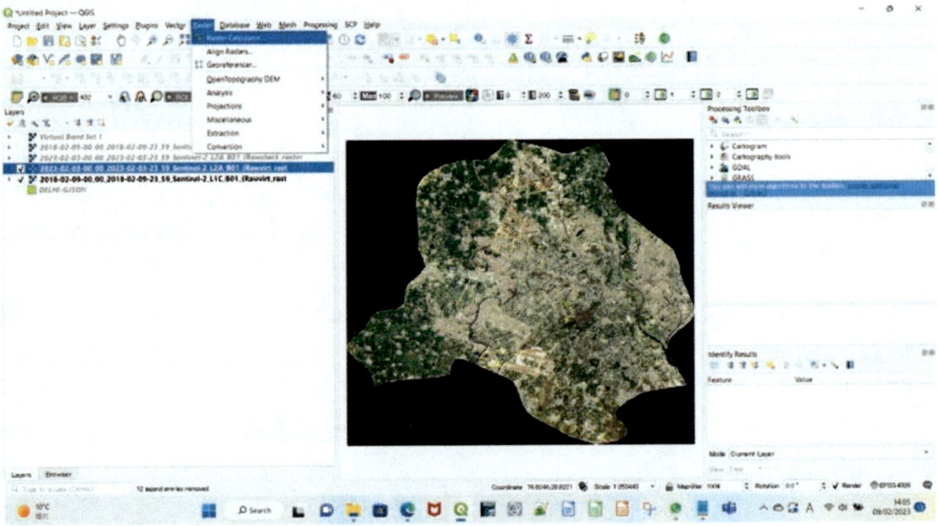

Fig. 10.14 Open raster calculator

After the window appear, input in the "Raster Calculator Expression" the following expression:

```
( "2018-02-09-00_00_2018-02-09-23_59_Sentinel-2_L1C_B01_(Rawvirt_
rast@8" - "2018-02-09-00_00_2018-02-09-23_59_Sentinel-2_L1C_B01_
(Rawvirt_rast@4" ) / ( "2018-02-09-00_00_2018-02-09-23_59_Sentinel-2_
L1C_B01_(Rawvirt_rast@8" +"2018-02-09-00_00_2018-02-09-23_59_Sentinel-2_
L1C_B01_(Rawvirt_rast@4" )
```

The name of file can be different, in principle we calculate using the following expression:

NDVI = (Band NIR-Band RED) / (Band NIR + Band RED).

Select the folder to save the result, give it the output layer name and click OK (Fig. 10.15).

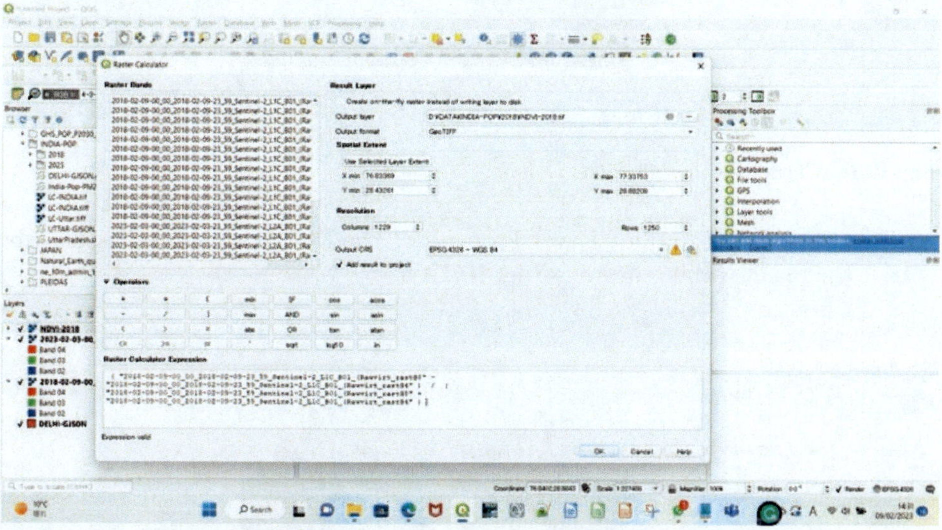

Fig. 10.15 Calculate the NDVI

The result will be visualized and greyscale, you need to change the symbology by double-clicking the layer and select “Symbology.” Change the “Render Type” to “Singleband pseudocolor” and click on the small drop-down arrow of “Color ramp” then select “Create new color ramp” > “Catalog: cpt-city.” Then scroll down slowly, choose “Greens” and “BuGn_09” (Fig. 10.16).

Fig. 10.16 Change the visualization

Then click "OK," in the "Layer Properties" window then click "Classify" the "OK." The result will be shown as Fig. 10.17.

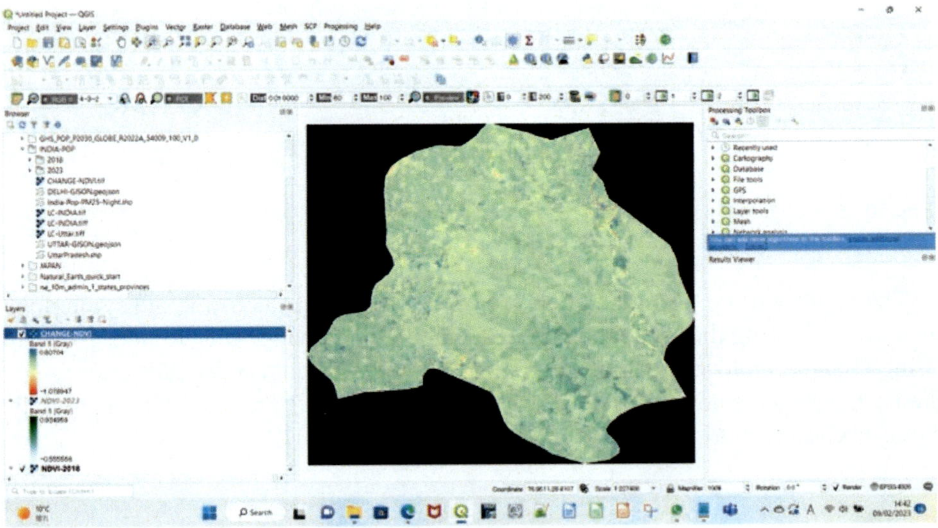

Fig. 10.17 The colorful result

Download again for different years from https://dataspace.copernicus.eu/. For this case we will download year 2023. Then calculate the NDVI for this year. Using "Raster Calculator" we will calculate the difference using following expression:

```
"NDVI-2023@1" - "NDVI-2018@1"
```

The file name can be different, in principle we want to calculate the different between year 2023 and 2018. Select the folder, give it name "CHANGE-NDVI" then click OK (Fig. 10.18).

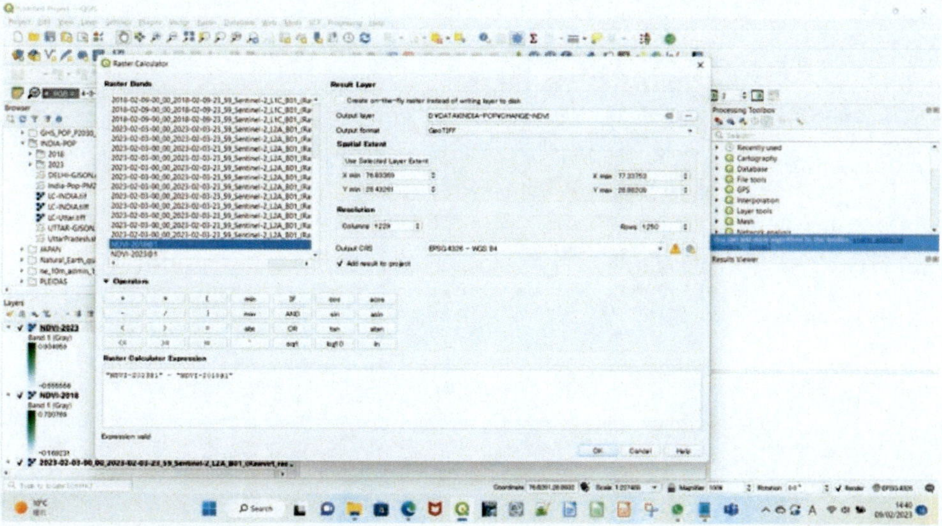

Fig. 10.18 Calculate the NDVI differences

If we zoom in to specific area, we will see the differences as shown in the following figure (Fig. 10.19).

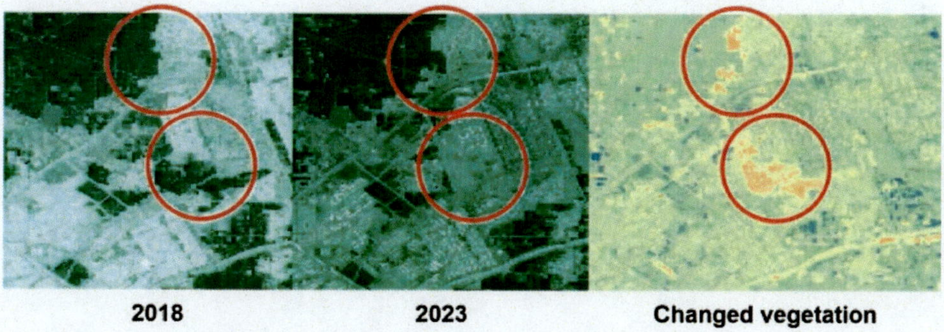

Fig. 10.19 NDVI changes over different time

10.1.1 Working with Model Designer

For next analysis, we need to develop a model using "Model Designer." The model is as shown in Fig. 10.20. In this model, we want to produce a vector data that consists of similar size of diamond shape. In this vector data we calculate the mean of vegetation density based on the input NDVI data using "Zonal statistics" algorithm.

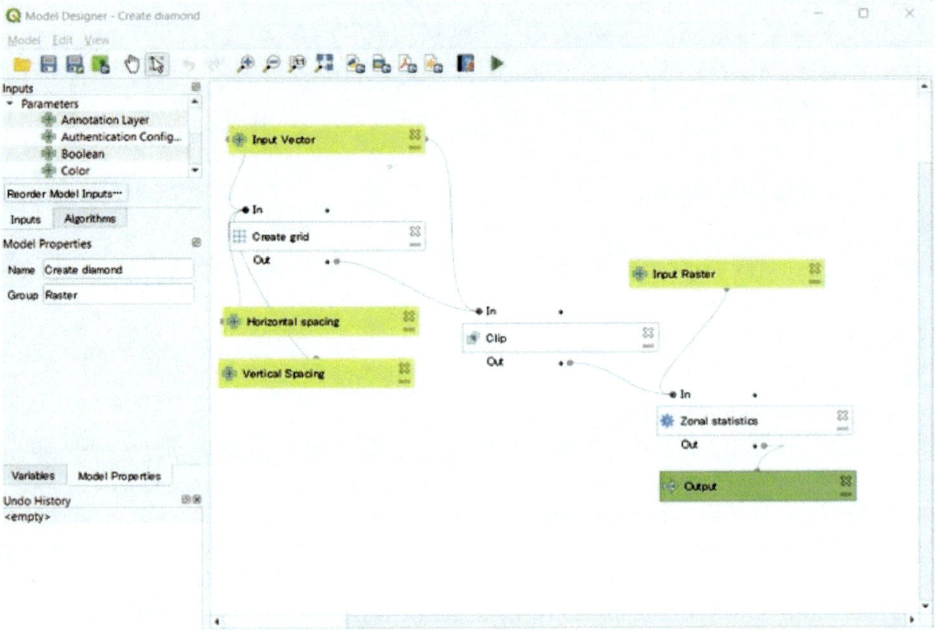

Fig. 10.20 Model designer to calculate the NDVI using diamond shape

To create a model, go to Processing > Graphical Modeler, and a new Model Designer window will open. To use vector as input layer, click on "Input," scroll down and double-click on "+Vector Layer." Type the description "Input," select "Any Geometry Type" as Geometry type, and check on "Mandatory" (Fig. 10.21).

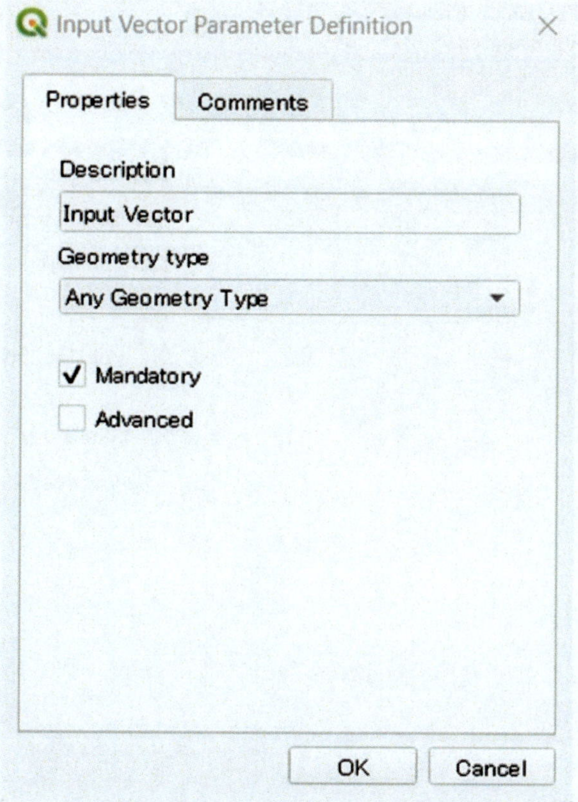

Fig. 10.21 Create input vector layer

Then scroll up look for "+String," double-click on it and input the "Horizontal spacing" in the Description, "1" in the Default value, and check on "Mandatory" (Fig. 10.22).

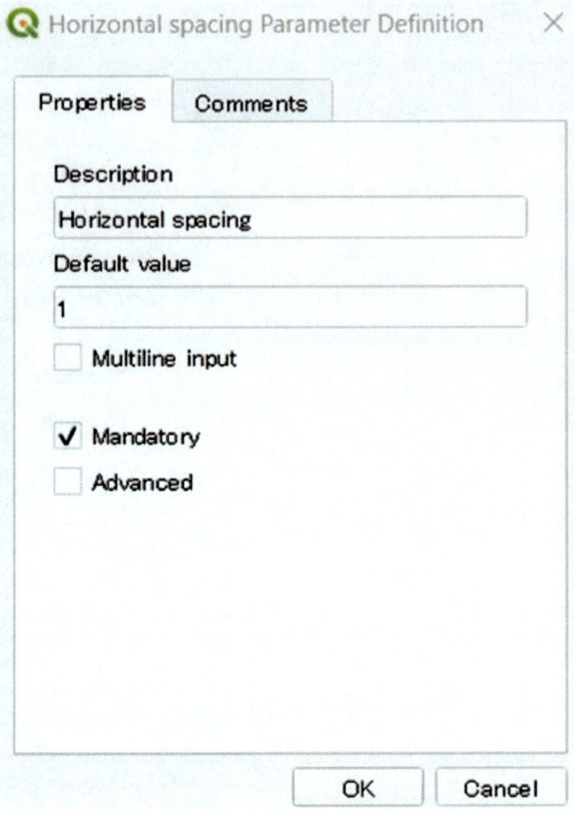

Fig. 10.22 Create horizontal spacing parameters definition

Double-click on "+String" again, and now input the "Vertical spacing" in the Description, "1" in the Default value, and check on "Mandatory" (Fig. 10.23).

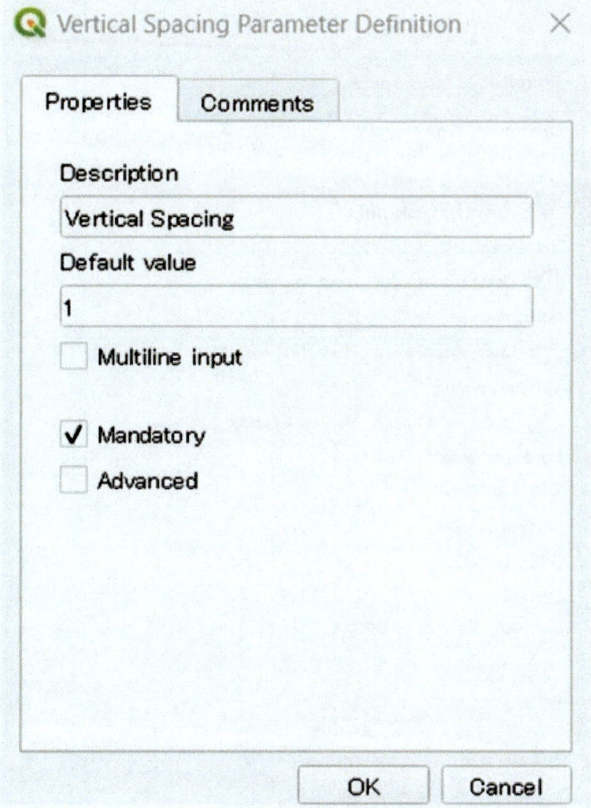

Fig. 10.23 Create vertical spacing parameters definition

To create grid, click on "Algorithms" and type "Create grid" in the search window, then double-click on it (Vector creation - Create grid). Type "Create grid" in the Description, select "Diamond (Polygon) under Grid Type. Under Grid extent select "Using model input" and select "Input Vector," while under Horizontal and Vertical extent select "Using model input" and select "Horizontal spacing" and "Vertical spacing," respectively. Then click OK (Fig. 10.24).

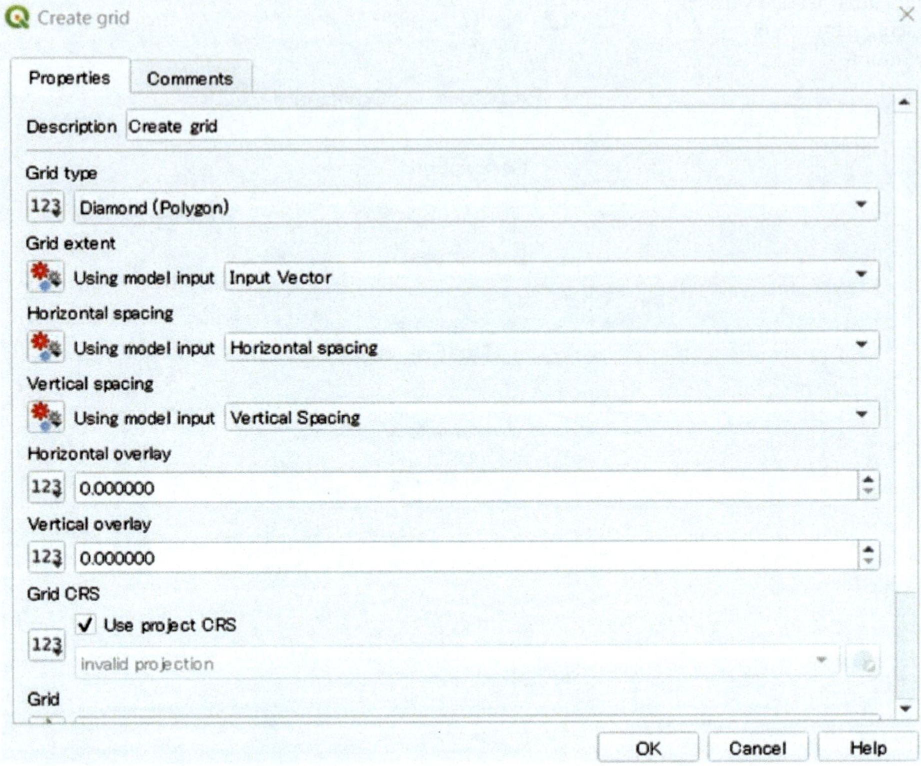

Fig. 10.24 Create diamond type grid

Next step is adding "Clip" (under Vector overlay > Clip), double-click and type "Clip" in the Description. In the Input layer select "Using algorithm output" and select "Grid" from algorithm "Create grid" and under Overlay layer select "Using model input" and select "Input Vector," then click OK (Fig. 10.25).

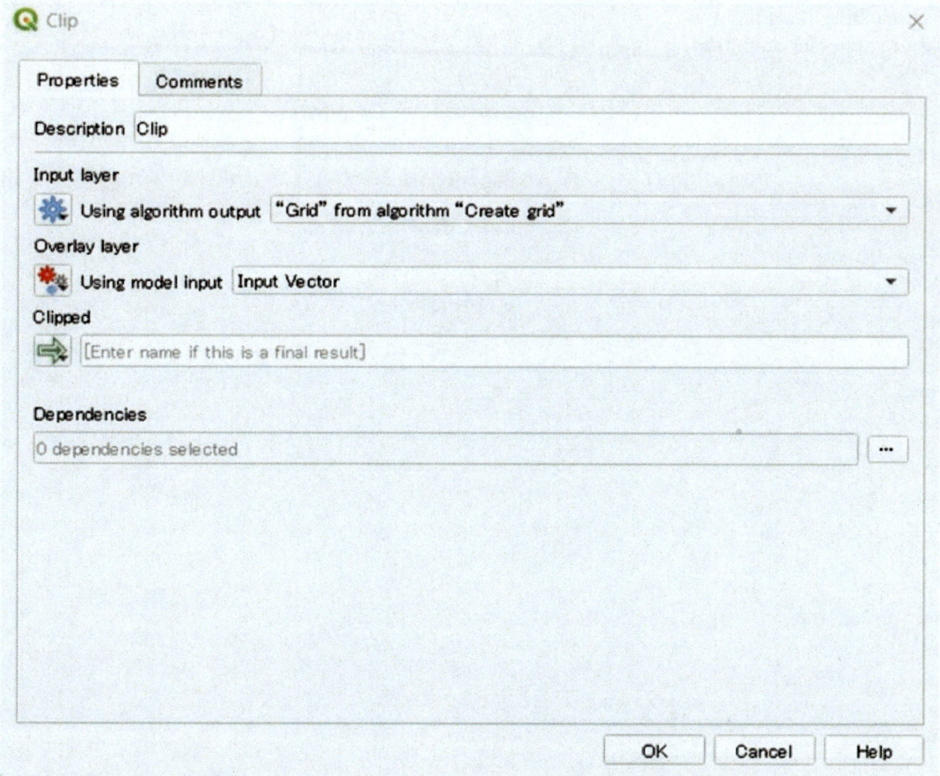

Fig. 10.25 Create clip function

Go back to "Input" and select "+Raster Layer." Type "Input Raster" in the Description and check on Mandatory then click OK (Fig. 10.26).

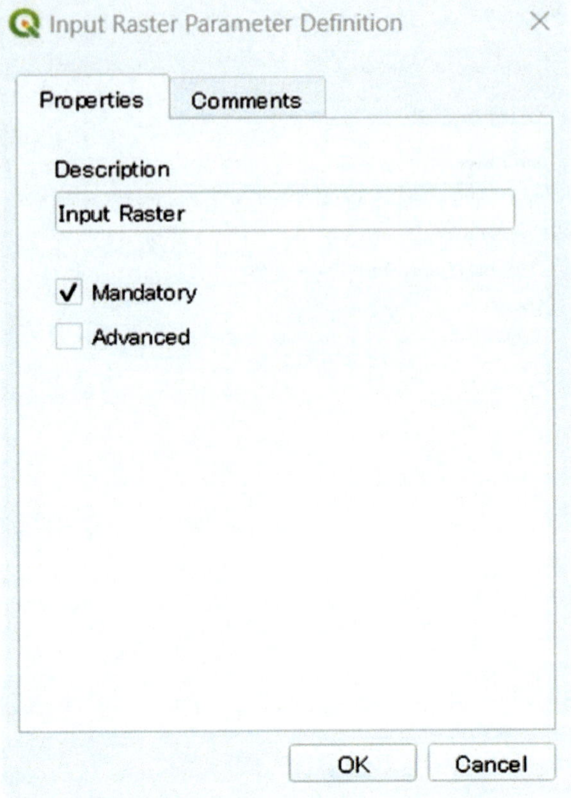

Fig. 10.26 Create input raster parameter definition

The last step is going back to "Algorithm" and type "Zonal" then select "Zonal statistics" (under Raster Analysis > Zonal statistics). Type "Zonal statistics" in the Description, select "Using algorithm output" under "Input Layer" and select "Clipped" from algorithm "Clip."

Under Raster layer select "Using model input" and select "Input Raster," while under Raster band type "1," under Output column prefix type "mean_", and under Statistics to calculate check on only "Mean" by click on three dots icon on the right. Finally, under the Zonal Statistics type "Output" as the final result (Fig. 10.27).

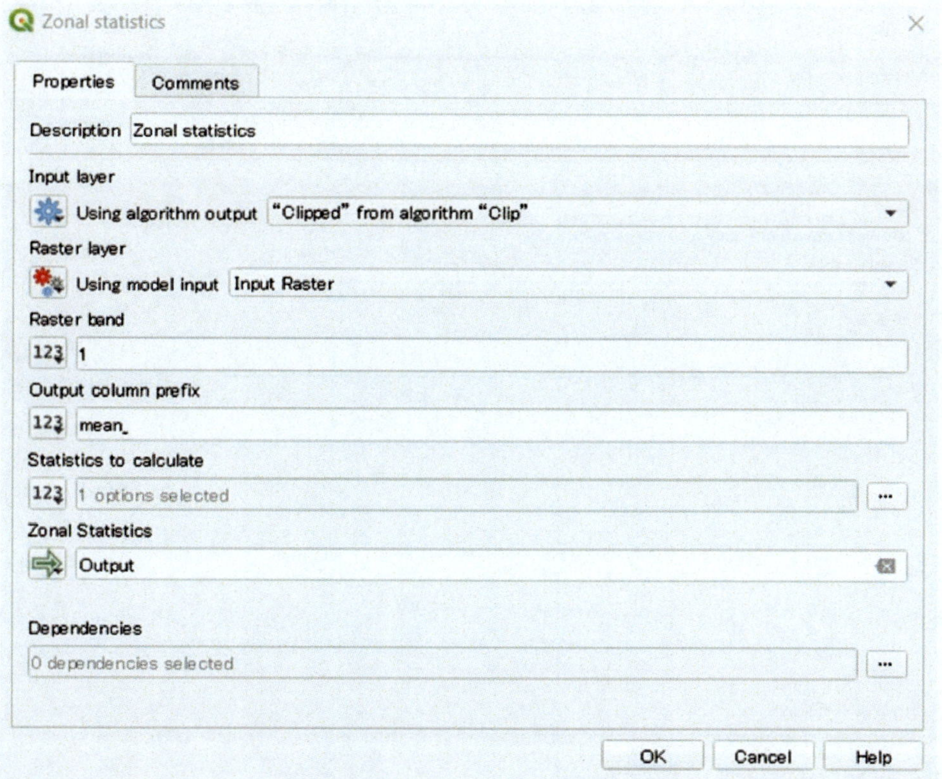

Fig. 10.27 Create zonal statistics function

To save the model give it the name under "Model Properties," type "Create Diamond" as Name and "Raster" as Group then go to Model > Save Model. To run the model, go to Model > Run Model.

If we run the model, the new window will be appeared as Fig. 10.28. Select the "Input" as vector boundary of New Delhi, then in the "Input Raster" select "CHANGE-NDVI." Make sure the "Horizontal and Vertical spacing" is same, in this case we input 0.01. Then click Run.

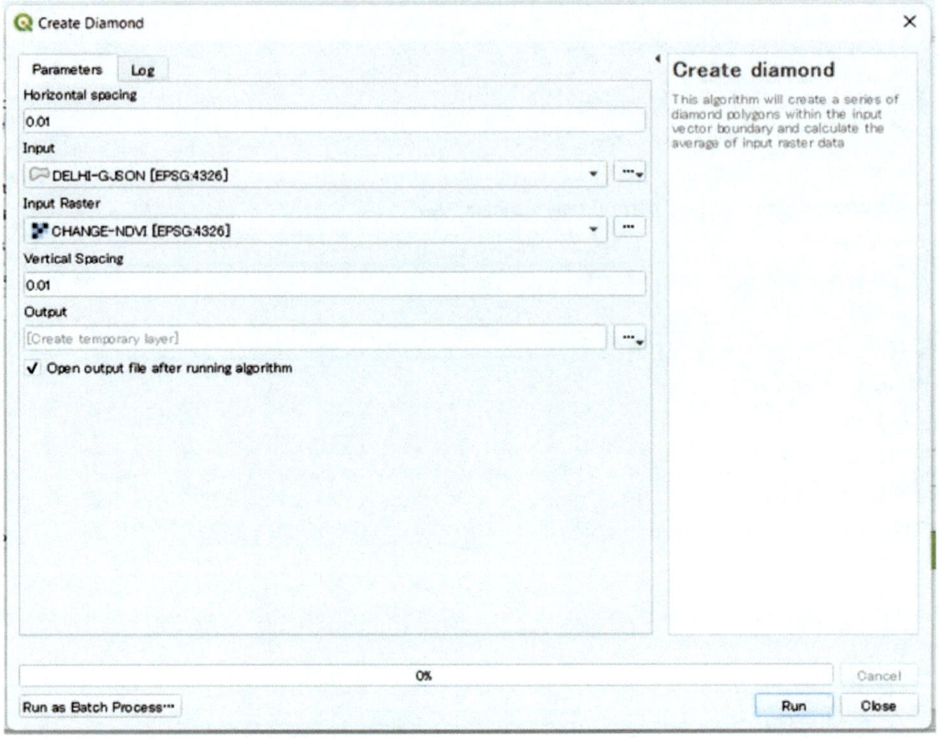

Fig. 10.28 Create diamond grid dialog box

The result will be a new vector data in diamond grid shape with average values of NDVI change from year 2018 to 2023. Double-click on the result layer the select "Symbology." Click on the icon  "Expression Dialog" next to "Value" (Fig. 10.29).

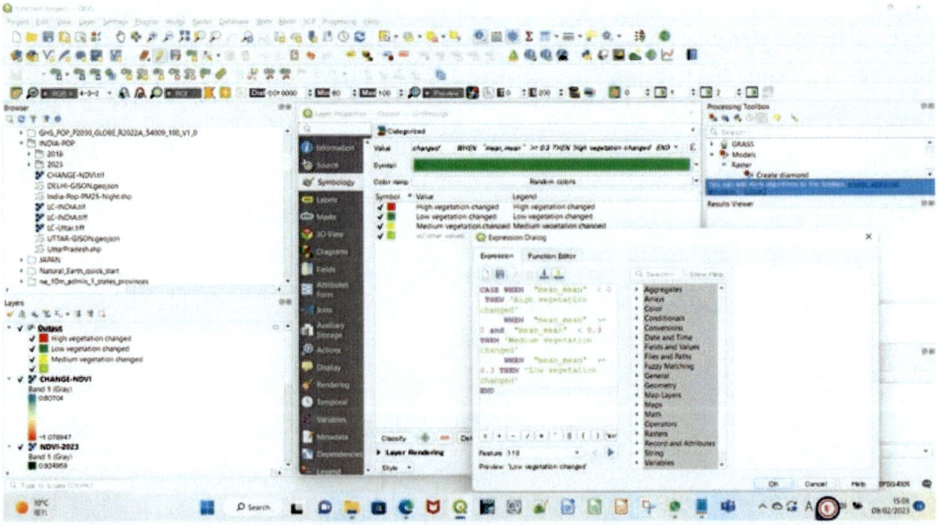

Fig. 10.29 Expression dialog box

Then input the following expression:

```
CASE WHEN  "mean_mean"  < 0 THEN 'High vegetation changed'
     WHEN  "mean_mean"  >= 0 and  "mean_mean"  < 0.3 THEN 'Medium
vegetation changed'
     WHEN  "mean_mean"  >= 0.3 THEN 'Low vegetation changed'
END
```

This expression means when the values are less than 0, it will be assigned as "High vegetation changed." While the values are more than or equal to 0 and less than 0.3, it will be assigned as "Medium vegetation changed." And when the values are more than or equal to 0.3, it will be assigned as "Low vegetation changed." Click OK.

You can change the color, "Red" for "High vegetation changed," "Green" for "Low vegetation changed," and "Yellow" for "Medium vegetation changed."

The final result is shown in Fig. 10.30. The result is easy to understand and locate the spatial distribution of vegetation changed within New Delhi.

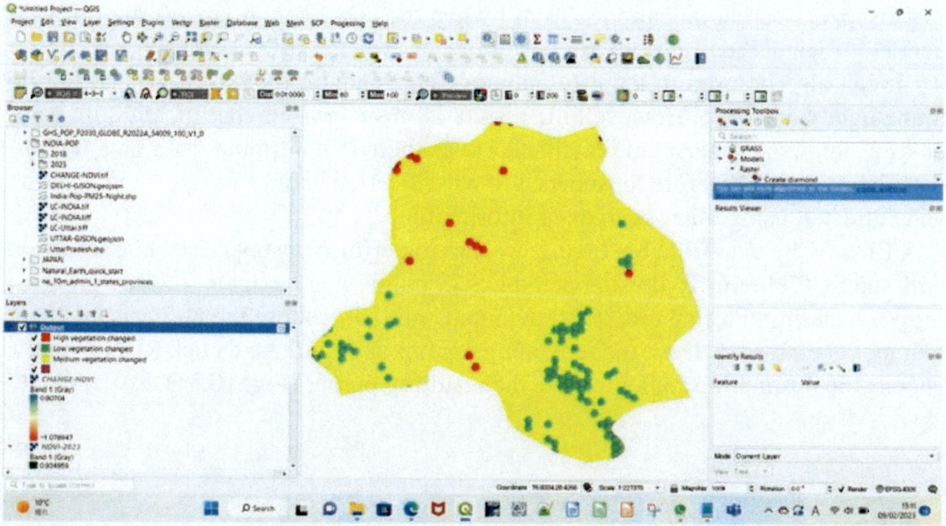

Fig. 10.30 Vegetation change overtime in New Delhi visualized using diamond grid

10.2 Land Use Land Cover Classification Using Machine Learning Algorithm

10.2.1 Unsupervised Classification

Unsupervised classification of satellite imagery is a type of image processing technique used to classify pixels in an image into distinct groups or classes without any prior knowledge of what those classes should be. Unlike supervised classification, where a predefined set of classes is used to label the data, unsupervised classification relies on algorithms to group pixels based on their spectral similarity, without any human input.

The process of unsupervised classification involves several steps:

- Image preprocessing: The first step is to preprocess the image to remove any unwanted noise and artifacts. This includes radiometric correction, atmospheric correction, and geometric correction.
- Segmentation: The image is then segmented into different regions or objects based on their similarity in terms of their spectral values. This is done using algorithms such as K-means clustering or hierarchical clustering.
- Feature extraction: Features such as texture, shape, and size are extracted from the segmented regions.
- Classification: The segmented regions are then classified into different classes based on their features. This can be done using clustering algorithms or by assigning each region to the class with the nearest mean value.
- Postprocessing: The final step involves refining the classification results and removing any misclassified pixels. This can be done manually or using automated techniques such as spatial filtering.

Unsupervised classification is useful when the classes in the image are not well defined or when the user does not have prior knowledge of what the classes should be. It can also be used to identify patterns and trends in the data that may not be immediately apparent. However, the results of unsupervised classification may not be as accurate as supervised classification, as there is no ground truth data to validate the results. Therefore, unsupervised classification should be used with caution and validated using other sources of information.

Utilizing the data from Sect. 10.1, we aim to perform unsupervised classification with the SCP plugin. To do so, open the SCP plugin and navigate to Band processing > Clustering. From there, select the input band set, choose the K-means method, Number of classes: 5, leave the remaining options as default, and click RUN. Finally, choose the destination folder to save the resulting output (Fig. 10.31).

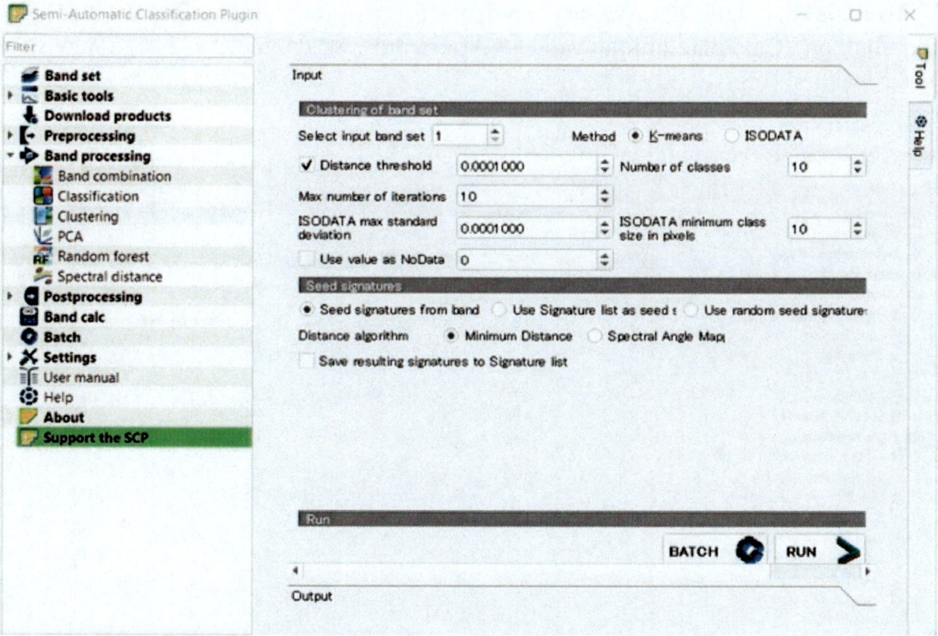

Fig. 10.31 Unsupervised classification using K-Means algorithm

The unsupervised classification result will be displayed in the main canvas in black and white color, after applying some symbology adjustments, the unsupervised classification result will be displayed on the main canvas as Fig. 10.32.

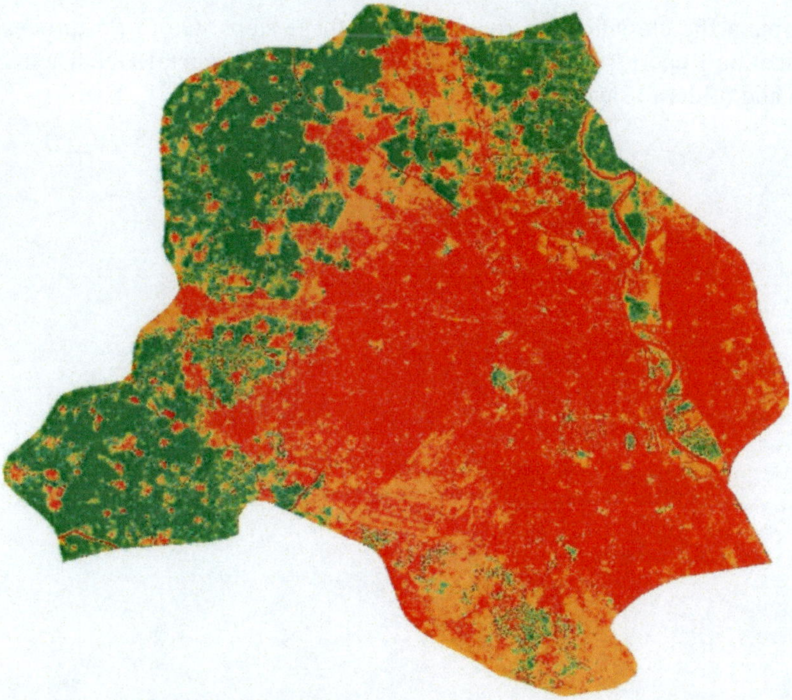

Fig. 10.32 Result of K-Means classifier

To reclassify data, if necessary, navigate to Postprocessing > Reclassification, and click on "Calculate unique values." Afterward, use the table to assign new values as needed (Fig. 10.33).

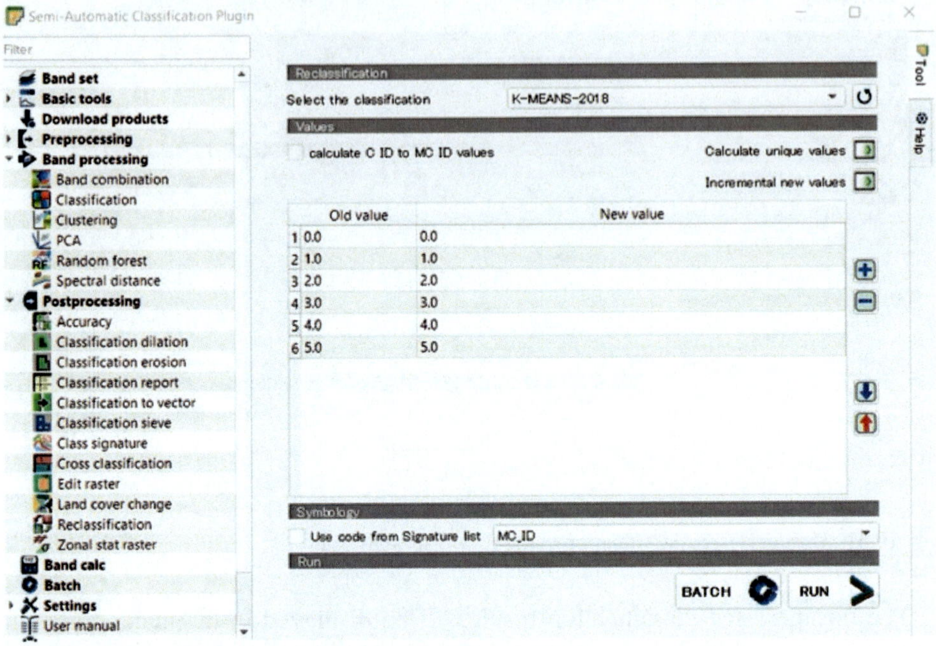

Fig. 10.33 Reclassify the result

To obtain the classification report, follow these steps: Go to Postprocessing > Classification Report, select the classification result, and click RUN. Then, choose a name and folder to save the report in (Fig. 10.34).

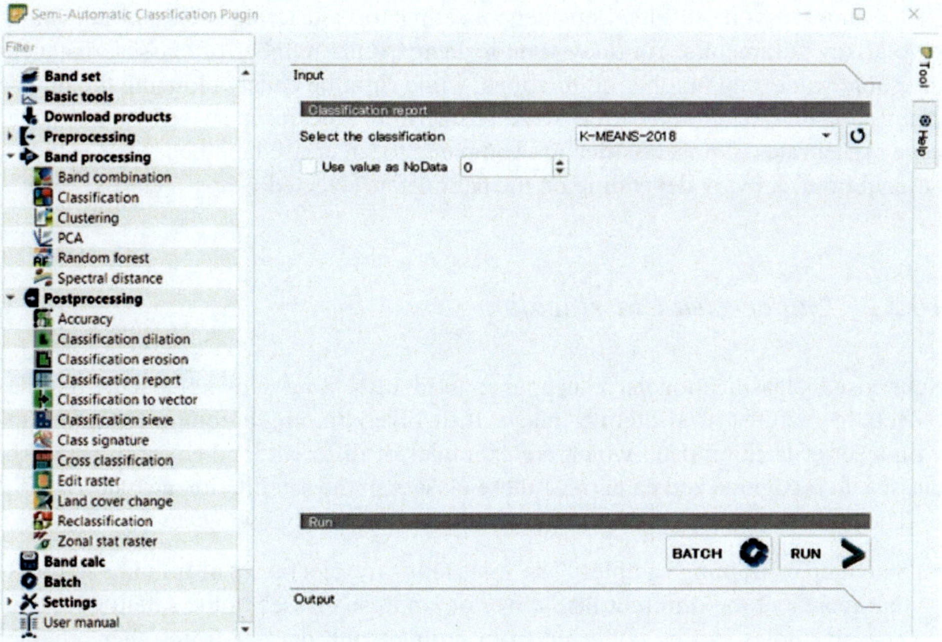

Fig. 10.34 Obtain the classification report

The following results will be displayed (Fig. 10.35), and you may disregard class 0 as it represents only the background of the raster data.

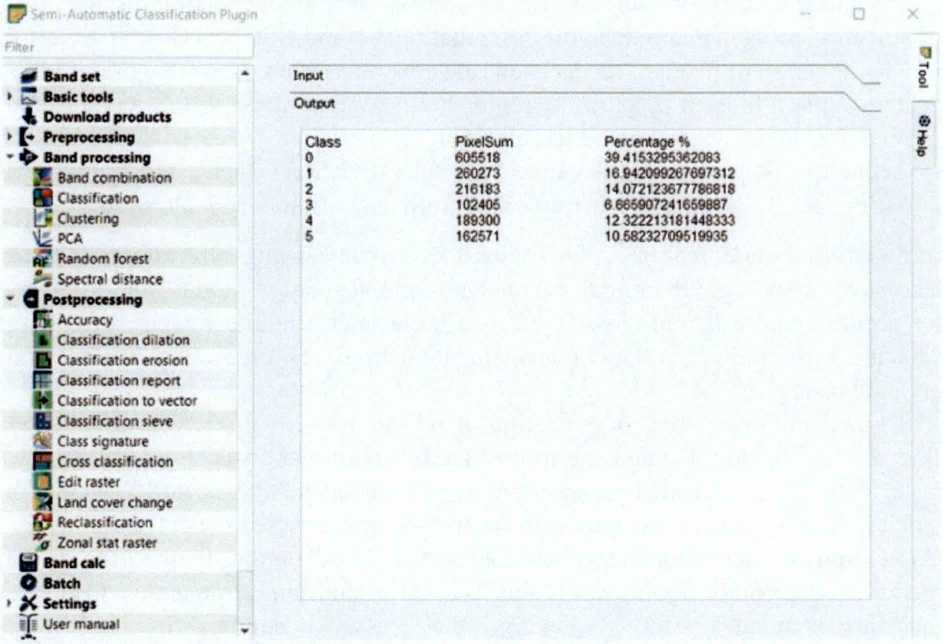

Fig. 10.35 Classification report (statistics)

To improve your results, consider repeating the clustering step and adjusting several key parameters. You may want to increase the number of classes, distance threshold value, and number of iterations, while simultaneously reducing the maximum standard deviation for ISODATA. Furthermore, the unsupervised methods can have higher rates of misclassification compared to supervised methods, and computational time can vary depending on the parameters selected.

10.2.2 Supervised Classification

Supervised classification is a technique used in remote sensing to classify the objects or features in a satellite image. It involves training a computer algorithm with a set of training data, which are examples of different land cover or land use classes, to recognize and categorize these classes in the satellite image.

The process of supervised classification involves the following steps:

- Selection of training samples: The first step is to select a set of training samples that represent the different land cover or land use classes in the satellite image. These samples can be collected from ground truth data or from other reliable sources.
- Classification algorithm: A classification algorithm is then applied to the training samples to create a decision rule that will be used to classify the pixels in the satellite image.
- Image segmentation: The satellite image is then segmented into smaller regions or segments based on their spectral similarity. This is done to simplify the classification process and reduce the computational complexity.
- Classification of pixels: The decision rule generated from the training samples is then applied to each pixel in the segmented image to classify it into one of the predefined land cover or land use classes.
- Accuracy assessment: The accuracy of the classification is then assessed by comparing the classified image with ground truth data or other reliable sources.

Supervised classification is widely used in remote sensing applications such as land cover mapping, urban land use analysis, and vegetation monitoring. It allows for accurate and efficient classification of large-scale satellite images, providing valuable information for land management, natural resource conservation, and urban planning.

To perform supervised classification, it is important to first create a band set (Fig. 10.36). To do this, navigate to the Bandset option and select the appropriate raster data. Then, add all necessary bands and select the consistent "Wavelength quick setting." If you are working with Sentinel-2 data, be sure to select "Sentinel-2." Check both "Create virtual raster" and "Create raster of" options before running the process. Additionally, double-check the "Band set definition" arrangement to ensure the arrangement before clicking the "Run" button. It must be B01-B02-B03-B04-B05-B06-B07-B08-B08A-B09-B10-B11-B12.

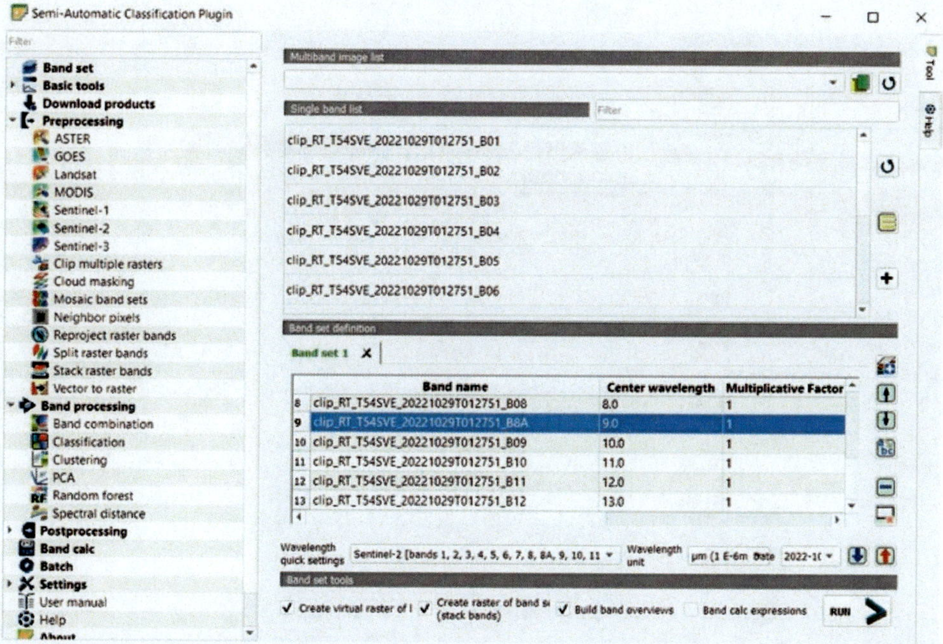

Fig. 10.36 Create a new band set

After finishing, the result will be displayed in the main canvas. Click on "Virtual Band Set 1" the look for this  and type 4–3-2 then hit enter; the main canvas now will show you new visualization of false color. You can change the visualization to 3–2-1 as well and hit enter button.

Next step is to create Region of Interest (ROI) as a new training input (Fig. 10.37). Click on icon  and select "Training input" then click on "Create a new training input" icon . Then give it a name and select the folder to save it.

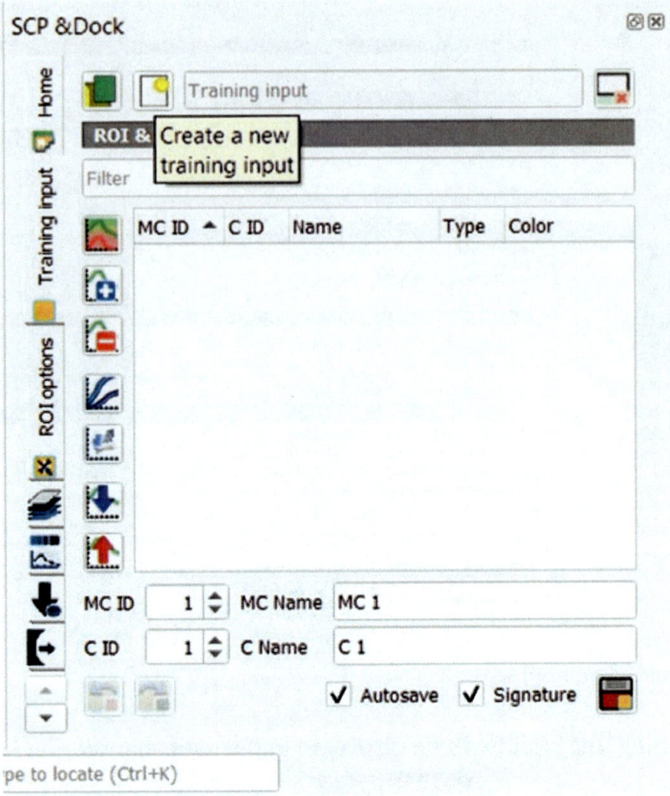

Fig. 10.37 Create ROI

To create the training input you have two options, "Create a ROI polygon" 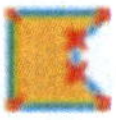 and "Activate ROI pointer" . The functionality is identical, but there is a distinction between the first and second features. The initial feature requires manual creation of the polygon, whereas the latter allows you to specify the location and with a simple click, generates the polygon automatically based on similar pixel values.

We will define four classes (MC Name) for this case: Urban (MC ID:1), Vegetation (MC ID:2), Bare land (MC ID:3), and Water body (MC ID:4). If necessary, we can also establish C ID and Name. It is important to remember to click the save icon button  after creating the training input (Fig. 10.38).

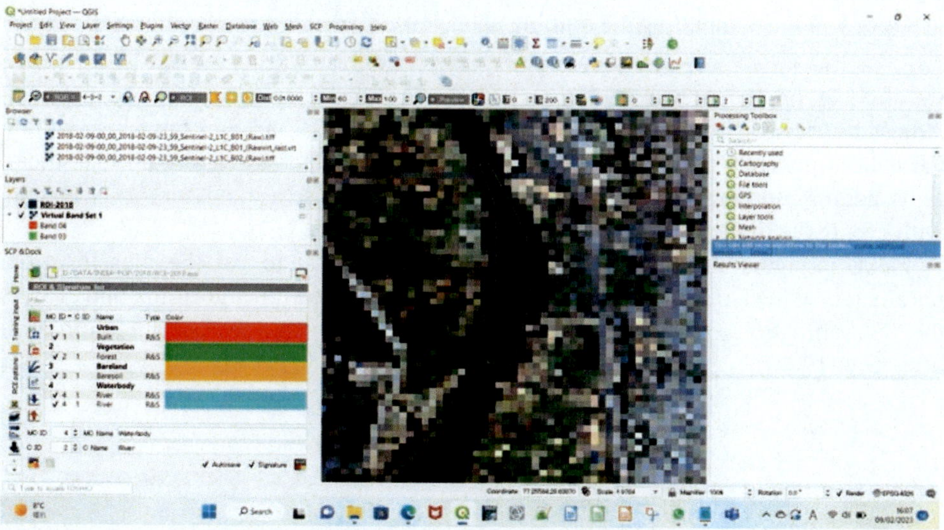

Fig. 10.38 Save the ROI

Additionally, SCP plugin has the capability to generate a spectral signature if necessary. To do so, simply select all classes and click on the "Add highlighted signatures to spectral signature plot" icon. The resulting plot will be displayed as shown in Fig. 10.39.

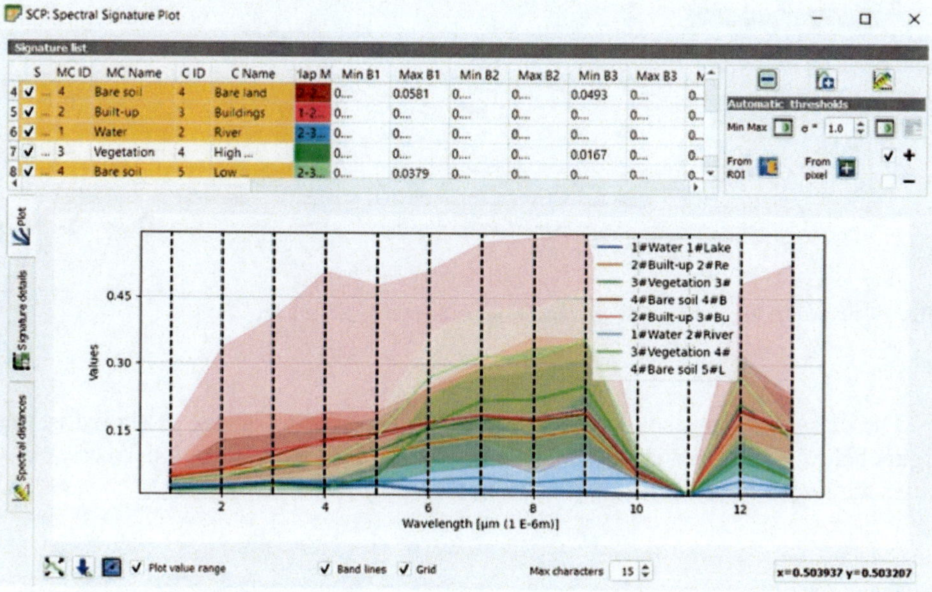

Fig. 10.39 Generate a spectral signature based on ROI

Once you finish inputting the training data, you can preview the classification results by clicking on the preview icon 

. Adjust the transparency level (T) and box size (S) as needed, and click anywhere within the data to generate the preview. Evaluate the preview and assess any misclassifications, adding more training data if necessary. Be sure to use a small, uniform polygon with evenly distributed and random pixels.

Next, navigate to Band processing > Classification to set the classification parameters. Check the MC ID, select the desired classification algorithm, and enable the "Classification report" option. Finally, click RUN to start the classification process (Fig. 10.40).

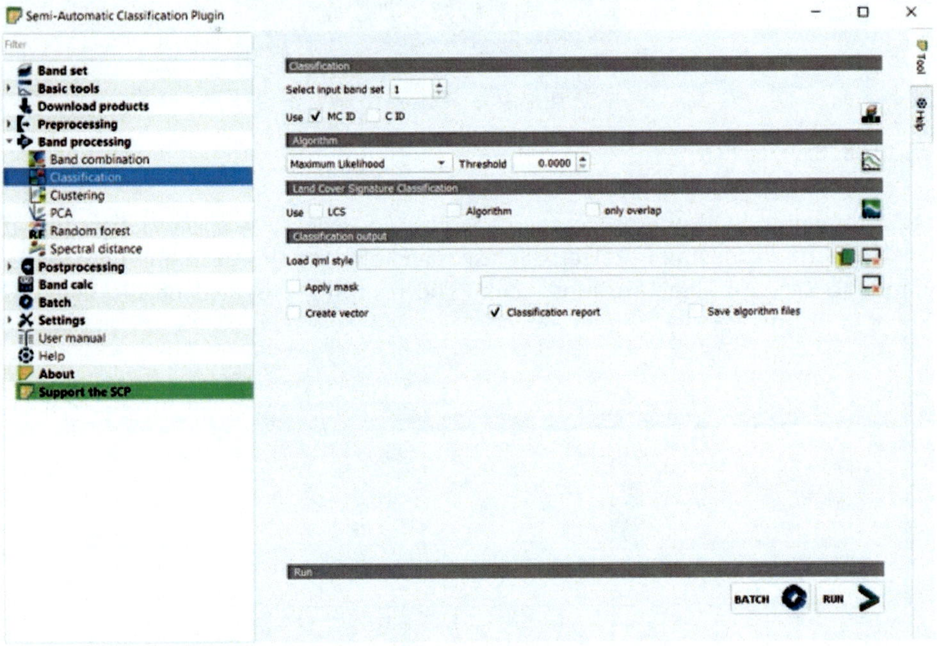

Fig. 10.40 Adjust the supervised classification parameters

The classification result will be displayed in the main canvas, as shown in the figure below. The supervised method has yielded superior results as it correctly classifies the river as one of the classification classes (Fig. 10.41).

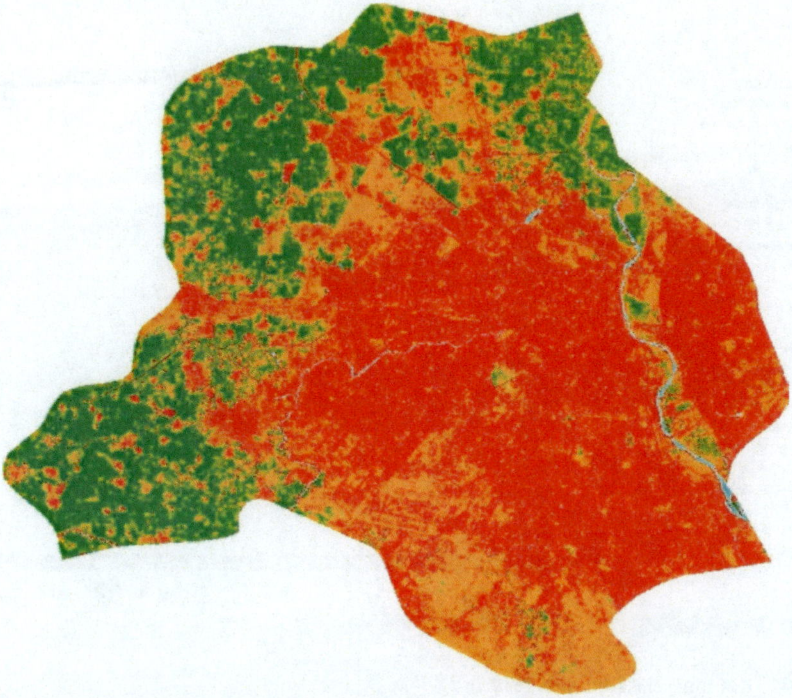

Fig. 10.41 Result of supervised Maximum Likelihood classifier

Supervised methods have certain limitation, including the need for high confidence and local knowledge on the part of the researcher. Additionally, creating more training datasets requires more time, and computational time is dependent on various factors such as the number of training datasets, the size of the study area, the pixel resolution of the satellite data, and the specifications of the computer used.

10.2.3 Refining Classification Result

The classification result is often affected by the *salt-and-pepper* issue, where numerous small pixels appear in the output. However, this can be addressed by refining the output through the "Postprocessing > Classification sieve" option (Fig. 10.42). By selecting the classification result as input and adjusting the "Size threshold" and "Pixel connection" parameters, we can achieve better results. For instance, setting the "Size threshold" to 4 will replace all patches smaller than 4 pixels with the value of the largest neighboring patch, while a "Pixel connection" value of 8 will consider diagonal pixels as connected.

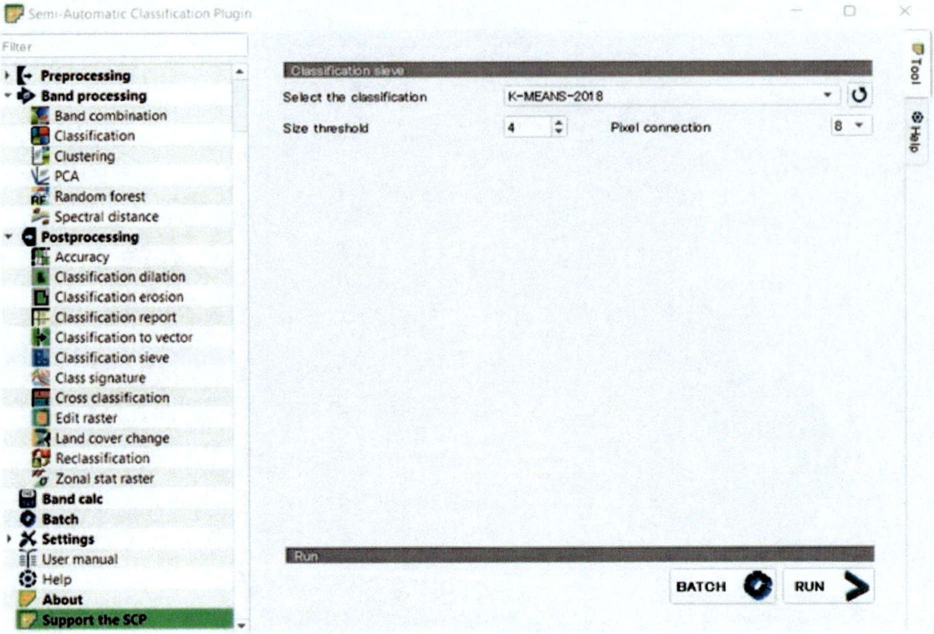

Fig. 10.42 Refining classification result

The result of before after refinement is shown in Fig. 10.43. Now we can see the small pixels disappear after refinement process.

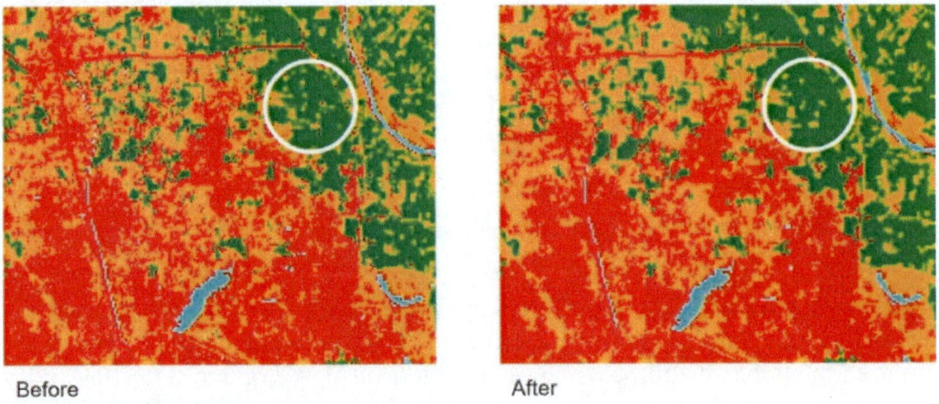

Fig. 10.43 Before and after refining classification result

10.2.4 Accuracy Assessment

To perform accuracy assessment, we can utilize the AcATaMa plugin in QGIS. This plugin offers extensive support for sampling, response design, and estimation within a design-based inference framework. Its primary objective is to equip users with the required tools to conform to standard guidance and best practices for sampling design, land category area estimation, land change, and map accuracy assessment.

To install AcATaMa, follow these simple steps: navigate to the Plugin menu, then select Manage and Install Plugin. Once the new window opens, type "acatama" in the search window. Next, select AcATaMa from the list of results and click on "Install Plugin" (Fig. 10.44).

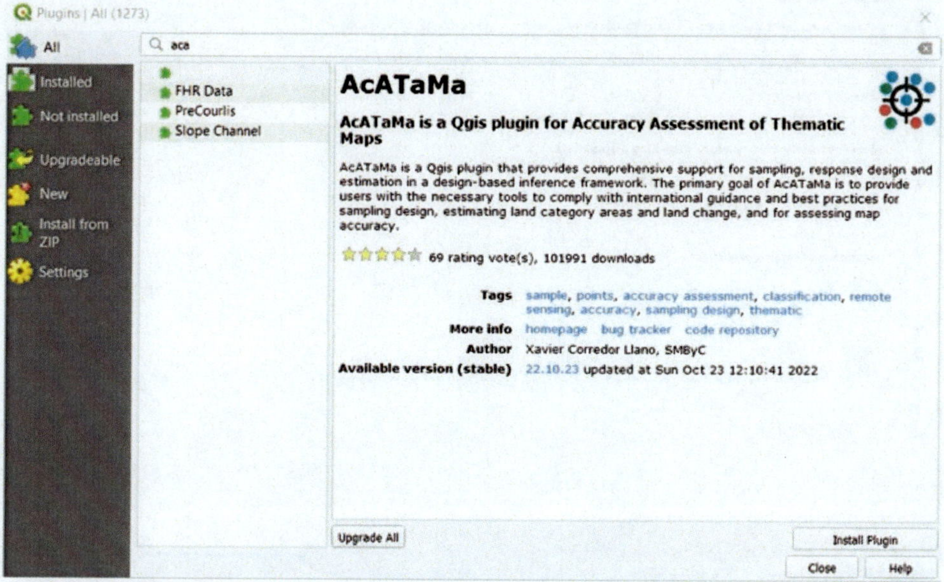

Fig. 10.44 Install acatama plugin

After completing the installation, search for the AcATaMa icon and click on it to open the window. If the window appears in the bottom-right panel of the QGIS window, drag it to the center and increase its size for better convenience. Next, choose the thematic map, which represents the classification results that will be evaluated for accuracy. Set the number of sample points to 50 and click on "Generate Samples." Finally, provide a name for the sample and select a folder to save it in (Fig. 10.45).

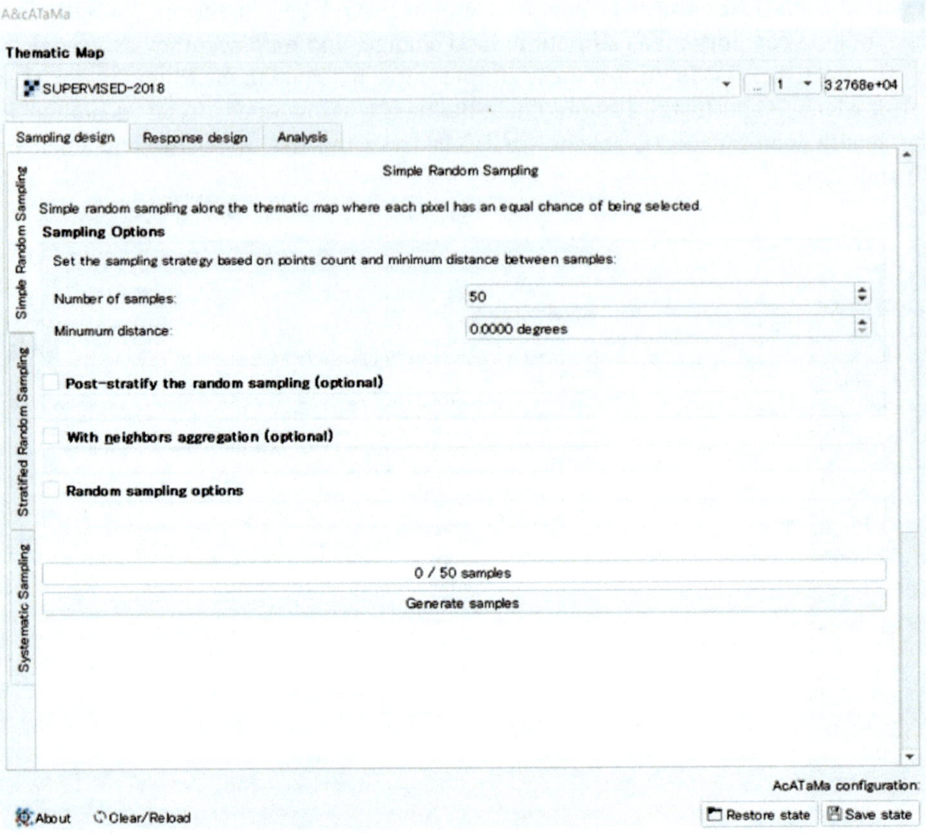

Fig. 10.45 Generate simple random sampling points

To display a Google Satellite based map, use the Quickmap Service plugin.

Go back to AcATaMa plugin and access the "Response design" tab and open the response design window, which will appear in a new window as shown in (Fig. 10.46). From the left panel, select "Google Satellite," and from the right panel, select the classification result.

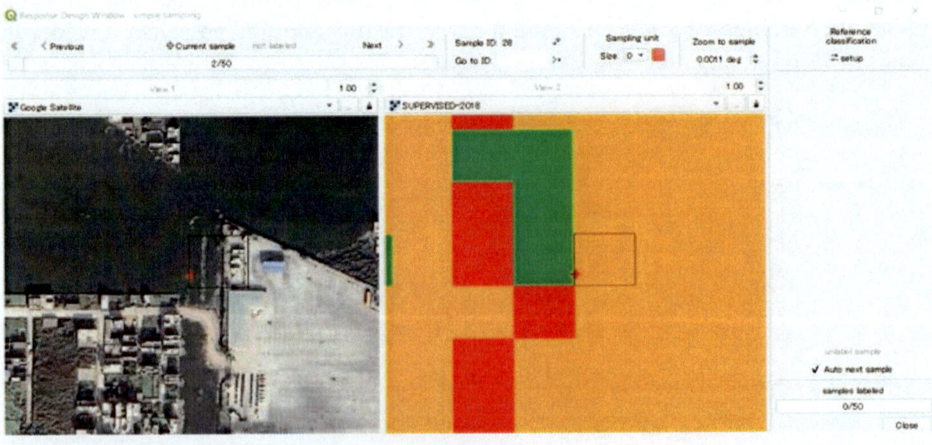

Fig. 10.46 Response design window dialog box

Click on the "Setup" icon 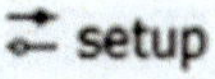

located under the "Reference Classification" section.

A window titled "Set the Labeling Buttons" will appear. Click on the first empty row under "Label Name" and enter a name for the label, such as "Urban." Click on the empty row under "Thematic Class" a new window will appear and select the corresponding class by clicking on "Select this" (Fig. 10.47).

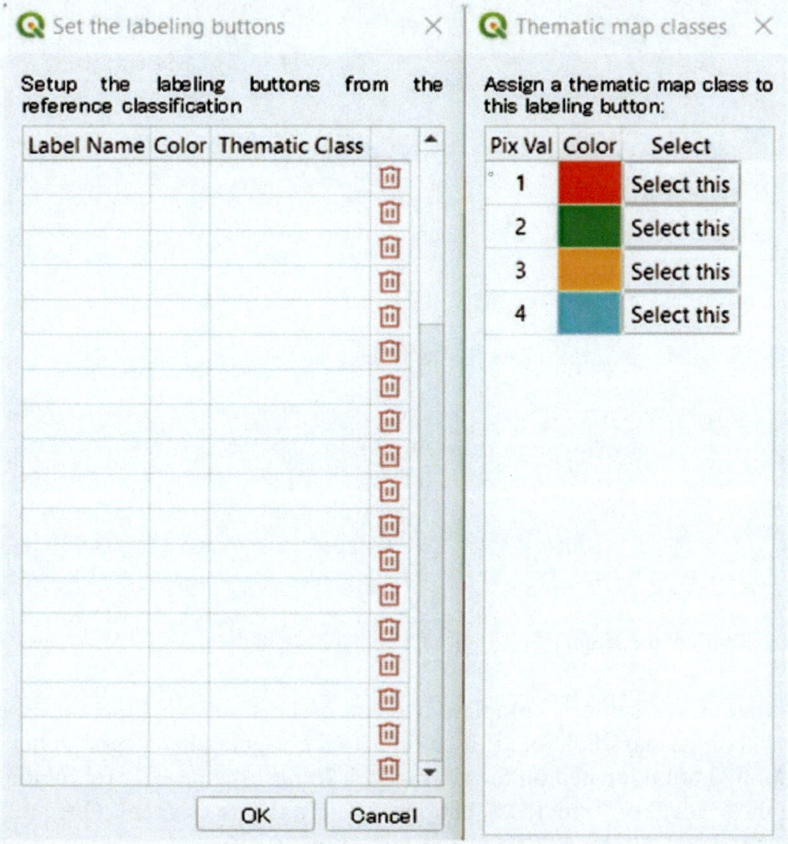

Fig. 10.47 Set the labeling buttons and corresponding thematic class

After assigning the label name and corresponding thematic class, the result will appear as shown in Fig. 10.48. The color and value may vary depending on the user's preference.

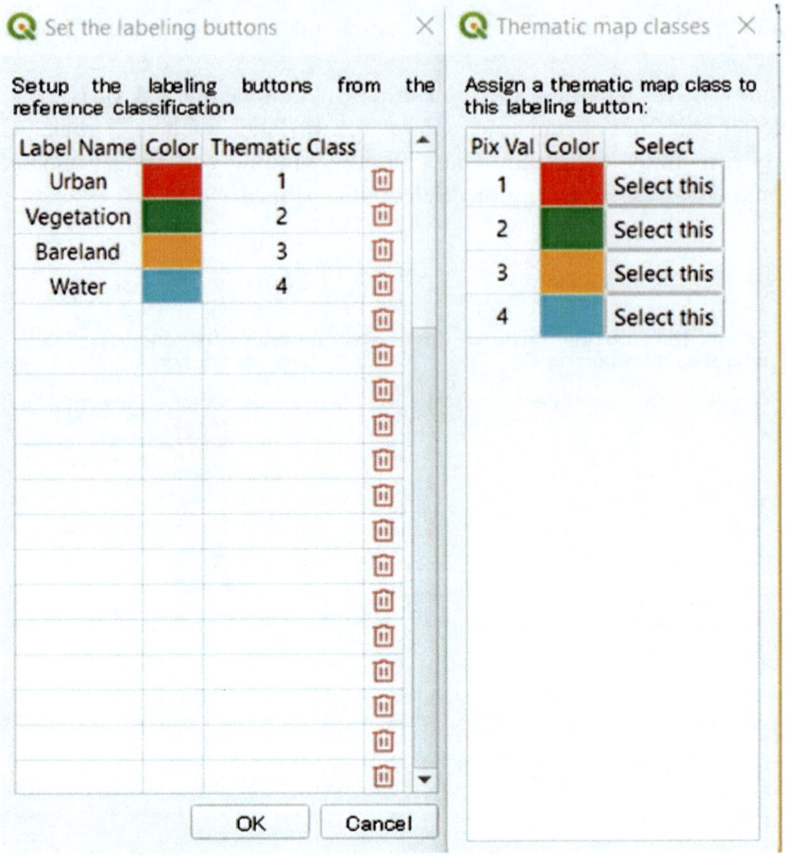

Fig. 10.48 Result of the setup

Now that all the setup is complete, we can proceed to select the corresponding point based on its classification. If it belongs to the vegetation category, click on the "Vegetation" button located on the right panel. Repeat the process for all the points and use the "Next" or "Previous" button to verify the assessment. Once all points have been labeled, click on "Close" to finish (Fig. 10.49).

Fig. 10.49 Select the corresponding point and class

After returning to the main window of AcATaMa, click on "Analysis." You should now see "Labeling completed 50/50" highlighted in green (Fig. 10.50), indicating that we can now compute the accuracy assessment.

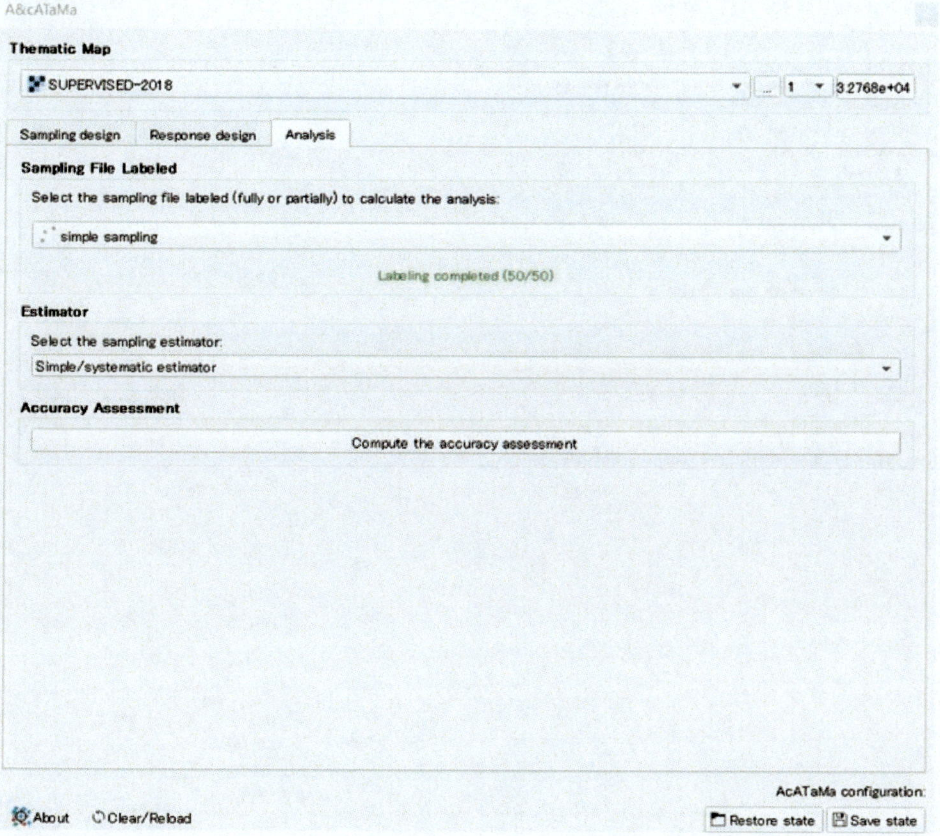

Fig. 10.50 Labeling completed

After clicking the "Compute the accuracy assessment" button, a new window will appear displaying the error matrix, accuracy, and class area adjustment. If necessary, you can change the area measurement unit to hectares. Finally, click on "Export to CSV" and save the file to your working folder (Fig. 10.51).

Fig. 10.51 Accuracy assessment result

10.3 Floods Mapping Using SAR

Floods mapping using Synthetic Aperture Radar (SAR) involves the use of SAR data to detect and map areas affected by floods. SAR is an active remote sensing technique that uses radar waves to generate images of Earth's surface. Unlike optical sensors, SAR can penetrate clouds, smoke, and other obstacles that may obstruct the view of the surface.

Here are the steps involved in floods mapping using SAR:

- Acquire SAR data: The first step is to acquire SAR data of the area of interest. SAR data can be obtained from satellite sensors such as Sentinel-1, ALOS-2, and RADARSAT-2.
- Preprocessing: The acquired SAR data undergoes preprocessing to remove noise, radiometric calibration, and geometric correction. The preprocessing stage is essential to improve the quality of the SAR images.
- Image analysis: The processed SAR images are analyzed to identify areas affected by floods. Flooded areas appear as dark regions in the SAR images due to the high backscatter of the radar waves from water surfaces.
- Image interpretation: The identified flooded areas are then interpreted to generate a flood map. The flood map shows the extent of flooding and the areas that are at high risk of flooding.
- Validation: The flood map is validated using ground-based data such as water level measurements and satellite imagery. The validation stage helps to improve the accuracy of the flood map.

SAR is a useful technique for floods mapping due to its ability to penetrate clouds and other obstacles. SAR data can be used to generate accurate and timely flood maps, which can aid in disaster management and mitigation.

In this case study we will map the extent of Pakistan's flood that was occurred on August 2022. Open the QGIS, copy and paste the following code into the python console:

```
import ee
from ee_plugin import Map

#Import country boundaries feature collection.
dataset = ee.FeatureCollection('USDOS/LSIB_SIMPLE/2017')

#Apply filter where country name equals Pakistan.
pakistan = dataset.filter(ee.Filter.eq('country_na', 'Pakistan'))

Map.centerObject(pakistan, 9);

#Filter the collection for the VV product before flood
collectionBef = ee.ImageCollection('COPERNICUS/S1_GRD')\
```

```
    .filter(ee.Filter.eq('instrumentMode', 'IW'))\
    .filter(ee.Filter.listContains('transmitterReceiverPolarisat
ion', 'VV'))\
    .filter(ee.Filter.eq('orbitProperties_pass', 'DESCENDING'))\
    .filterDate('2022-06-01','2022-06-30')\
    .filterBounds(pakistan)\
    .select(['VV']);

#Filter the collection for the VV product after flood
collectionAft = ee.ImageCollection('COPERNICUS/S1_GRD')\
    .filter(ee.Filter.eq('instrumentMode', 'IW'))\
    .filter(ee.Filter.listContains('transmitterReceiverPolarisat
ion', 'VV'))\
    .filter(ee.Filter.eq('orbitProperties_pass', 'DESCENDING'))\
    .filterDate('2022-08-01','2022-08-31')\
    .filterBounds(pakistan)\
    .select(['VV']);

before = collectionBef.median().clip(pakistan);
Map.addLayer(before, {'min': -14, 'max': -7}, 'Before',0);

after = collectionAft.median().clip(pakistan);
Map.addLayer(after, {'min': -14, 'max': -7}, 'After',0);

#Reduce noise (radar speckle) with smoothing
smoothing_radius = 100;
before_filtered = before.focal_mean(smoothing_radius, 'circle',
'meters');
after_filtered = after.focal_mean(smoothing_radius, 'circle',
'meters');

Map.addLayer(before_filtered, {'min':-25, 'max':0}, 'Before_filtered',0);
Map.addLayer(after_filtered, {'min':-25, 'max':0}, 'After_filtered',0);

#Calculate the difference between the before and after images
difference = after_filtered.divide(before_filtered);
Map.addLayer(difference,{'min':0, 'max':2},"Difference Layer",0);

#Apply the predefined difference-threshold and create the flood extent
mask
threshold = 1.5;
difference_binary = difference.gt(threshold);

#Refine flood result using additional datasets
```

```
#Include JRC layer on surface water seasonality to mask flood pixels
from areas of "permanent" water (where there is water > 10 months
of the year)
swater = ee.Image('JRC/GSW1_0/GlobalSurfaceWater').select(
'seasonality');
swater_mask = swater.gte(10).updateMask(swater.gte(10));

#Flooded layer where perennial water bodies (water > 10 mo/yr) is
assigned a 0 value
flooded_mask = difference_binary.where(swater_mask,0);

#final flooded area without pixels in perennial waterbodies
flooded = flooded_mask.updateMask(flooded_mask);

#Compute connectivity of pixels to eliminate those connected to 10
or fewer neighbours
#This operation reduces noise of the flood extent product
connections = flooded.connectedPixelCount();
flooded = flooded.updateMask(connections.gte(10));

#Mask out areas with more than 5 percent slope using a Digital
Elevation Model
DEM = ee.Image('WWF/HydroSHEDS/03VFDEM');
terrain = ee.Algorithms.Terrain(DEM);
slope = terrain.select('slope');
flooded = flooded.updateMask(slope.lt(5));

#Display the result
Map.addLayer(flooded,{'palette':"0000FF"},'Flooded areas');
```

For better visualization add base map using QuickMapServices plugin, install it and add Google Map Road. The final result will be shown as Fig. 10.52.

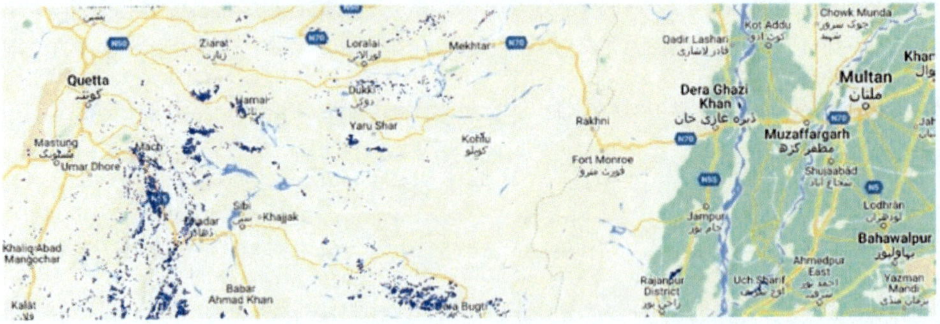

Fig. 10.52 Flooding in Pakistan as detected using Sentinel-1

Pakistan, a nation positioned at the confluence of the Indian subcontinent, Central Asia, and the Arabian Sea stands vulnerable to recurrent and devastating floods as a direct consequence of its distinctive location and intricate geography. Nestled amidst these diverse geographical influences, Pakistan is home to an intricate network of colossal rivers, most notably the mighty Indus, Jhelum, and Chenab rivers. These waterways, while vital for irrigation and sustenance, also become menacing catalysts during the monsoon season, when torrential downpours mercilessly pummel the region. The deluge of heavy monsoon rains engorges the rivers, leading to their unforgiving overflow and subsequent submergence of adjacent areas, leaving behind a trail of destruction and human suffering.

Glossary and Abbreviations

Acronym or concept	Meaning or description
ALOS	Advanced Land Observing Satellite is a series of satellites developed by the Japan Aerospace Exploration Agency (JAXA) for Earth observation purposes. The ALOS program consists of several satellites, including ALOS-1 (also known as "Daichi") and ALOS-2 (also known as "Daichi-2").
ASTER	Advanced Spaceborne Thermal Emission and Reflection Radiometer is an instrument onboard NASA's Terra satellite. It captures high-resolution imagery in multiple spectral bands, including visible, near-infrared, and thermal infrared. ASTER data is primarily used for mapping and monitoring Earth's land surface temperature, elevation, and geology.
Aspect	It refers to the compass direction that a slope or terrain surface faces. It is an essential component of terrain analysis and represents the orientation of a land surface relative to the cardinal directions (north, south, east, and west). The aspect is measured in degrees clockwise from north, ranging from 0° to 360°
Classification	In the context of LULC, it involves assigning specific land cover categories to each pixel in an image. It is typically performed using machine learning algorithms, where the algorithm is trained on a set of labeled training data that consists of pixels with known land cover categories. Once trained, the algorithm can classify unseen pixels in the image based on their spectral characteristics and contextual information.
Clip	It refers to the operation of extracting a subset or a portion of a spatial dataset based on a defined boundary or extent. When clipping a DEM, you define a boundary or a polygon that outlines the area of interest, and the operation retains only the elevation data within that boundary. The resulting clipped DEM will contain elevation values only for the specified area, while the rest of the dataset outside the boundary will be excluded.
Clustering	It refers to data analysis technique that involves grouping similar data points together based on their similarities and dissimilarities. In the context of LULC, clustering algorithms can be used to identify patterns and group similar land cover features together.

F. Ramdani, *Exploring the Earth with QGIS*, Springer Remote Sensing/Photogrammetry, https://doi.org/10.1007/978-3-031-46042-5

Acronym or concept	Meaning or description
Copernicus	Copernicus is a European Union Earth observation program aimed at providing accurate and timely information for environmental monitoring and civil security. It encompasses a range of satellite missions and ground-based observations.
CRS	Coordinate Reference System is a framework used to define and interpret geographic locations on Earth's surface. It provides a standardized way to represent and store spatial data. A CRS consists of two main components: a coordinate system and a datum.
DEM	Digital Elevation Model is a digital representation of Earth's terrain, typically in a raster format, where each cell in the raster grid represents the elevation or height above sea level of a specific location on Earth's surface. DEMs are commonly used in various applications such as topographic mapping, hydrological analysis, and 3D visualization.
DSM	Digital Surface Model is a digital representation of Earth's surface, including both natural and man-made features. It represents the elevations of the surface objects, such as buildings, trees, mountains, and other terrain features, including the bare ground.
Elevation profile extraction	It refers to the process of obtaining the elevation data along a specific path or route. It involves determining the varying elevation values at different points along the path, typically represented as a graph or a series of data points.
EPSG	European Petroleum Survey Group is an organization that originally defined a standardized set of coordinate systems and projections for use in geospatial applications. The EPSG Geodetic Parameter Dataset (EPSG dataset) was developed and maintained by this group.
GEE	Google Earth Engine is a cloud-based platform developed by Google for analyzing and visualizing geospatial data. Google Earth Engine provides access to a vast amount of satellite imagery and geospatial datasets, along with tools for data analysis and remote sensing applications.
GIS	Geographic Information System refers to a computer-based system designed to capture, store, manipulate, analyze, and present spatial or geographic data. GIS integrates various types of data, such as maps, satellite imagery, aerial photographs, and tabular data, into a digital environment where relationships and patterns can be identified, visualized, and understood.
Hillshade	It refers to a technique used to represent the relief or topography of an area on a 2D map or image. It creates a visual representation of how light and shadows interact with the terrain, simulating the effect of sunlight on the landscape.
Landsat	Landsat program is a series of Earth observation satellites (Landsat 4, 5, 7, 8, and 9) jointly operated by NASA (National Aeronautics and Space Administration) and the USGS (United States Geological Survey). The satellites capture images of Earth's surface, primarily in visible and infrared wavelengths. Landsat data is widely used for monitoring land cover changes, assessing vegetation health, mapping urban areas, and studying natural resources.
LIDAR	Light detection and ranging is a remote sensing technology that uses laser pulses to measure distances and create detailed three-dimensional representations of Earth's surface. It works by emitting laser beams and measuring the time it takes for the light to bounce back after hitting an object. LIDAR is commonly used in applications such as topographic mapping, autonomous vehicles, forestry, and environmental monitoring.

Acronym or concept	Meaning or description
LST	Land surface temperature refers to the temperature of Earth's land surface, including the soil, vegetation, and man-made structures. It is a measure of the thermal energy emitted by the land surface and is usually expressed in degrees Celsius or Kelvin.
LULC	Land use and land cover refers to the categorization and mapping of different types of land uses and land covers in a specific area, typically derived from remotely sensed data such as satellite imagery. LULC classification involves assigning specific classes to different areas or pixels within an image, representing different land use and land cover categories such as forests, urban areas, agricultural fields, water bodies, etc.
Map Algebra	A technique used in GIS to perform mathematical operations on spatial data layers or maps. It involves combining or manipulating raster or vector data layers using various mathematical operators, such as addition, subtraction, multiplication, and division. Map Algebra allows for the integration of different datasets and the generation of new spatial information.
MODIS	Moderate resolution imaging Spectroradiometer is an instrument aboard NASA's Terra and Aqua satellites. It provides high-resolution imagery and collects data across a wide range of the electromagnetic spectrum, from visible to thermal infrared wavelengths. MODIS is known for its global coverage, daily revisits, and its ability to capture data on a variety of environmental parameters, such as land surface temperature, vegetation indices, cloud cover, and ocean color.
NDVI	Normalized Difference Vegetation Index is a commonly used vegetation index in remote sensing. It measures the health and density of vegetation by analyzing the difference between the near-infrared (NIR) and red light reflected by plants. NDVI values range from −1 to 1, with higher values indicating healthier and denser vegetation.
OBIA	Object-based image analysis is an approach to image analysis and classification that considers the spatial context and relationships between neighboring pixels. Instead of classifying individual pixels, OBIA groups pixels into meaningful objects or segments based on their spectral, spatial, and contextual characteristics. These objects are then classified into different land cover classes.
Pixel	A pixel is the smallest unit of an image. In the context of remote sensing and LULC mapping, pixels represent individual cells or picture elements within an image, each containing spectral information captured by the remote sensing sensor. Each pixel is characterized by its spectral values, such as reflectance or radiance, which can be used for LULC classification.
Plugin	Plugin is an add-on or extension that provides additional functionality to the software. Plugins are designed to enhance the core features of QGIS and allow users to customize their workflows or add specific tools and capabilities. They are created by the QGIS community, including developers and users, and can be installed and enabled within the QGIS application.
Point sampling	It refers to the process of extracting the elevation profile by selecting a specific path or route and sampling elevation values along that path. The path can be defined using a set of coordinates or by tracing a line on a map or a digital terrain model.

Acronym or concept	Meaning or description
QGIS	Quantum GIS is a free and open-source GIS software that allows users to visualize, analyze, and manage geospatial data. It is widely used by professionals in various fields such as geography, environmental science, urban planning, cartography, biology, economy, engineering, law, etc.
Random forest	A supervised machine learning algorithm that is used for both classification and regression tasks. It is an ensemble learning method that combines multiple decision trees to make predictions.
Resample	It refers to the process of changing the resolution or spatial sampling of a raster dataset. When resampling a DEM, the cell size or pixel size of the raster is modified to a new value. Resampling can be done to increase or decrease the resolution of the DEM, depending on the specific requirements of the analysis or application.
Salt-and-pepper	Salt-and-pepper issue typically refers to a visual artifact or problem that can occur in digital images. It manifests as randomly occurring black and white pixels scattered throughout the image, resembling grains of salt and pepper. These pixels are essentially digital noise that can distort the quality and clarity of the image.
SAR	Synthetic Aperture Radar is another remote sensing technology that uses radio waves to generate high-resolution images of Earth's surface. Unlike optical sensors that rely on sunlight, SAR can operate day and night and in all weather conditions. It works by transmitting microwave signals towards Earth's surface and recording the signals reflected back.
SAVI	Soil-Adjusted Vegetation Index is an index used in remote sensing to quantify and analyze vegetation cover in relation to soil background. It is an improvement over the Normalized Difference Vegetation Index (NDVI) that accounts for variations in soil reflectance. SAVI is particularly useful in areas with dense vegetation or where soil brightness significantly affects the accuracy of vegetation analysis.
Sentinel	Sentinel is a series of satellite missions developed by the European Space Agency (ESA) as part of the Copernicus program, which is Europe's flagship Earth observation program. The Sentinel satellites (Sentinel-1, 2, 3, 5p, and 6) are equipped with various sensors to monitor different aspects of the Earth, including land, ocean, atmosphere, and climate.
Slope	It refers to the steepness or incline of the land surface. It is a measure of how much elevation changes over a given horizontal distance. The lower the slope value, the flatter the terrain; the higher the slope value, the steeper the terrain.
SST	Sea surface temperature refers to the temperature of the uppermost layer of the ocean, typically the top few meters. It represents the heat content of the ocean surface and is an essential parameter for understanding and predicting weather patterns, climate variations, and ocean dynamics.
Supervised learning	A machine learning approach where a model is trained on labeled data, meaning the data has input features and corresponding target variables or categories. The model learns from the labeled examples and tries to generalize the patterns in the data to make predictions or classify new, unseen data.

Acronym or concept	Meaning or description
SVM	Support Vector Machine is a supervised machine learning algorithm used for classification and regression analysis. The objectives of SVM are to find a hyperplane in an n-dimensional feature space that best separates the data points belonging to different classes. In a binary classification scenario, the algorithm determines an optimal decision boundary that maximizes the margin between the two classes. This decision boundary is defined by a subset of the training samples called support vectors, which are the data points closest to the hyperplane.
Training and testing data	The training data consist of labeled pixels or samples with known land cover categories. These samples are used to train a classification algorithm to learn the relationship between the spectral characteristics and the corresponding land cover classes. The testing data, on the other hand, are a separate set of unlabeled pixels or samples that are used to evaluate the performance of the trained algorithm. the algorithm's accuracy and generalization ability are assessed by comparing the predicted land cover classes with the true classes of the testing data.
Unsupervised learning	A machine learning approach in which a model is trained on unlabeled data without any predefined target variables or categories. The goal of unsupervised learning is to discover patterns, relationships, or structures in the data.
Viewshed	It refers to the area or terrain visible from a specific vantage point or observer location. It is commonly used in GIS and remote sensing to analyze and visualize the visibility of an area from a given point. the viewshed analysis takes into account factors such as terrain elevation, obstructions like buildings or vegetation, and the observer's viewing distance or height.
Watershed	An area of land from which all the water drains into a common outlet, such as a river, lake, or ocean. It is also known as a catchment or drainage basin. Watersheds are important units for studying and managing water resources. They are delineated based on topographic features such as ridges and peaks that define the boundaries of where water flows.
Zonal statistics	It refers to a method or technique used in spatial analysis to calculate statistical summaries or measurements within predefined zones or regions. These zones can be defined by polygons, such as administrative boundaries, land parcels, or other spatial units.

Bibliography

N. Baghdadi, C. Mallet, and M. Zribi, *QGIS in Remote Sensing Set. Volume 2, QGIS and Applications in Agriculture and Forest* (Wiley, 2018)

A. Cutts, *QGIS Quick Start Guide: A beginner's Guide to Getting Started with QGIS 3.4* (Packt Publishing, 2019)

A. Graser, G.N. Peterson, *QGIS Map Design*, 2nd edn. (Locate Press LLC, 2020)

P. Holloway, *Understanding GIS Through Sustainable Development Goals: Case Studies with QGIS* (CRC Press Taylor & Francis Group, 2023)

S. Islam, S. Miles, K. Menke Jr., R.S. Pirelli, J. van Hoesen, *Mastering Geospatial Development with QGIS 3.x: An In-depth Guide to Becoming Proficient in Spatial Data Analysis Using QGIS 3.4 and 3.6 with Python*, 3rd edn. (Packt Publishing, 2019)

K. Menke, *Discover QGIS 3.x: A Workbook for Classroom or Independent Study*, 2nd edn. (Locate Press LLC, 2022)

F. Ramdani, P. Setiani, Data of satellite observation for environmental assessment before and during COVID-19 pandemic in part of Indonesia using the cloud-computing platform. Geosci. Data J. **9**(2), 304–314 (2022). https://doi.org/10.1002/gdj3.144

P. Setiani, L.A. Devianto, F. Ramdani, Rapid estimation of CO_2 emissions from forest fire events using cloud-based computation of Google earth engine. Environ. Monit. Assess. **193**(10), 1–13 (2021). https://doi.org/10.1007/s10661-021-09460-w

H. van der Kwast, K. Menke, *QGIS for Hydrological Applications: Recipes for Catchment Hydrology and Water Management* (Locate Press LLC, 2020)

F. Ramdani, *Exploring the Earth with QGIS*, Springer Remote Sensing/Photogrammetry, https://doi.org/10.1007/978-3-031-46042-5

Index

F. Ramdani, *Exploring the Earth with QGIS*, Springer Remote Sensing/Photogrammetry, https://doi.org/10.1007/978-3-031-46042-5